Current and Emerging Trends in Cyber Operations

Palgrave Macmillan's Studies in Cybercrime and Cybersecurity

This book series addresses the urgent need to advance knowledge in the fields of cybercrime and cybersecurity. Because the exponential expansion of computer technologies and use of the Internet have greatly increased the access by criminals to people, institutions, and businesses around the globe, the series will be international in scope. It provides a home for cutting-edge long-form research. Further, the series seeks to spur conversation about how traditional criminological theories apply to the online environment. The series welcomes contributions from early career researchers as well as established scholars on a range of topics in the cybercrime and cybersecurity fields.

Series Editors:

MARIE-HELEN MARAS is Associate Professor and Deputy Chair for Security at the Department of Security, Fire, and Emergency Management at John Jay College of Criminal Justice, USA.

THOMAS J. HOLT is Associate Professor in the School of Criminal Justice at Michigan State University, USA.

Titles include:

Amitai Etzioni
PRIVACY IN CYBER AGE
Policy and Practice

Frederic Lemiux (editor)
CURRENT AND EMERGING TRENDS IN CYBER OPERATIONS
Policy, Strategy and Practice

Current and Emerging Trends in Cyber Operations

Policy, Strategy, and Practice

Edited by

Frederic Lemieux
George Washington University, USA

palgrave
macmillan

First published 2015 by
PALGRAVE MACMILLAN

Palgrave Macmillan in the UK is an imprint of Macmillan Publishers Limited, registered in England, company number 785998, of Houndmills, Basingstoke, Hampshire RG21 6XS.

Palgrave Macmillan in the US is a division of St Martin's Press LLC, 175 Fifth Avenue, New York, NY 10010.

Palgrave is the global academic imprint of the above companies and has companies and representatives throughout the world.

Palgrave® and Macmillan® are registered trademarks in the United States, the United Kingdom, Europe and other countries.

ISBN: 978–1–137–45554–3

This book is printed on paper suitable for recycling and made from fully managed and sustained forest sources. Logging, pulping and manufacturing processes are expected to conform to the environmental regulations of the country of origin.

A catalogue record for this book is available from the British Library.

A catalog record for this book is available from the Library of Congress.

Library of Congress Cataloging-in-Publication Data

Current and emerging trends in cyber operations : policy, strategy and practice / [edited by] Frederic Lemieux, George Washington University, USA.
pages cm
Includes bibliographical references and index.
ISBN 978–1–137–45554–3 (hardback)
1. Cyberterrorism. 2. National security. I. Lemieux, Frédéric.
HV6773.15.C97C87 2015
355.4—dc23

2015013078

'The Chinese use two brush strokes to write the word "crisis." One brush stroke stands for danger; the other for opportunity. In a crisis, be aware of the danger – but recognize the opportunity.'

John F. Kennedy, Speech in Indianapolis, Indiana, 12 April 1959

Contents

Acknowledgments

First of all, I would like to express my deepest gratitude to all contributors who made this project possible. I want to thank all of the authors for the originality and the high quality of the work they produced. This book represents a major contribution to the field of international police cooperation, and achieving this objective in a short period of time was a heavy demand. Also, I was delighted to work with the publishing team at Palgrave Macmillan. The confidence they had in the project and their judicious advice was instrumental to the realization of the book. I am deeply indebted to my special assistant, Melinda Hull, who worked hard on the revision and editing of the chapters. Thank you, Melinda, for having been flexible and reliable and for offering excellent suggestions throughout the editing process. Finally, I am grateful to my wife, Alterra Hetzel, who is always supportive of my work, for her dedication to our family.

Notes on Contributors

Editor

Frederic Lemieux is a full professor and program director of the Master's Degree in Homeland Security and the Master's Degree in Cybersecurity Strategy and Information Management, and he is co-director of the Bachelor's Degree in Cybersecurity at the George Washington University. He also co-founded the George Washington University Cyber Academy. Frederic Lemieux has published several books, book chapters, and journal articles in the field of homeland security, international police cooperation, and cyber security.

Contributors

Lieutenant Colonel **Scott D. Applegate** is a career military officer with more than 22 years of experience. He is the operations chief of Defensive Cyberspace Operations at the US Army Cyber Command. He is a published author and a past speaker at a number of conferences, including Hacker Halted and the International Conference on Cyber Conflict. His research interests include information assurance, cyber conflict, cyber militias, security metrics, and mobile device security.

Zal Azmi is the chief executive officer for Nexus Solutions LLC. He offers more than 30 years of leadership experience and demonstrated success in the development, nurturing, program management, performance management, organizational maturity, and operational integration of advanced technology systems and solutions to meet a variety of enterprise modernization needs. In his last ten years in the government, he served as the chief information officer (CIO) for United States Attorneys (2000–04) and the FBI (2004–09), where he established the CIO organization's information assurance and cybersecurity programs.

Jeffrey Bardin is the chief intelligence strategist at Treadstone 71. He has more than 25 years of experience in the fields of IT and information security, risk management and assurance, cyber intelligence and counterintelligence. Since 1982, Jeffrey Bardin has worked in leadership positions in organizations such as General Electric, Lockheed Martin, and Marriott International. He also served as the security manager for the Centers for Medicare and Medicaid (LMIT), chief security officer for Hanover Insurance, the chief information security officer for Investors Bank & Trust, and the director of the Office of Risk Management for EMC.

Andrew Colarik is an independent consultant, author, researcher, and inventor of information security technologies. He serves as a senior lecturer in the Department of Computer Science at Auckland University, New Zealand. He has published multiple security books and publications in the areas of cyber terrorism, information warfare, and cyber security. His primary research areas are the security impact of the global information infrastructure on businesses, governments, and individuals; the technology impact on social, political, legal, and economic structures in society; and the design and implementation of secure communication systems.

Jack Goldsmith is the Henry L. Shattuck Professor at Harvard Law School, where he teaches and writes about national security law, presidential power, cyber security, international law, Internet law, foreign relations law, and conflict of laws. Before coming to Harvard, Professor Goldsmith served as assistant attorney general at the Office of Legal Counsel from 2003–04, and as special counsel to the Department of Defense from 2002–03.

Wayne Harrop is the director of the Centre for Disaster Management at the University of Coventry in the UK. He has developed a hybrid career as an academic and practitioner, winning three international industry accolades and contributing to funded research projects worldwide. Mr. Harrop is part of a national cyber-security advisory cell led by the Bank of England. Mr. Harrop co-directs the International Risk, Resilience and Response Centre (a UK–US transatlantic partnership), which has successfully delivered prime ministers' funded projects on international dimensions of 'urban crisis' (providing international briefings on homeland security, disaster impacts, national infrastructure, and cyber security).

Roland Heickerö is an adjunct professor at KTH Royal Institute of Technology in Sweden. He was previously working at the Swedish National Defense College (SNDC). His research examines different aspects of information warfare and cyber threats and their effects at the security policy level as well as on social and technical systems levels. Between 2003 and 2012, he was deputy research director at the Swedish Defense Research Agency (FOI) in charge of cyber defense research.

Lech Janczewski is Associate Professor of Information Systems and Operations Management (Business School) at the University of Auckland, New Zealand. He has over 35 years of experience in information technology. He was the managing director of the largest IBM installation in Poland and the project manager of the first computing center in the Niger State of Nigeria. His area of research includes management of IS resources with a special emphasis on data security and information systems investments. He contributes to a project aimed at developing a tool handling distributed denial of service attacks.

Jyh-An Lee is an assistant professor in the Faculty of Law at the Chinese University of Hong Kong. His research interests include intellectual property, information law, and Internet law. Dr. Lee holds a JSD from Stanford Law School and an LLM from Harvard Law School. He is the author of two books: *Coding a Free Society: Open Source Strategies for Policymakers* (VDM Verlag Müller Press, 2007) and *Nonprofit Organizations and the Intellectual Commons* (Edward Elgar, 2012). Before starting his academic career, Jyh-An Lee was a practicing lawyer in Taiwan specializing in technology and business transactions.

Ashley Matteson serves as a steering group member and cybersecurity advisor to the International Risk, Resilience and Response Centre, chaired jointly by Coventry University in the UK and Texas A&M University Engineering Extension in the US. Mr. Matteson has completed and become certified in all foundation and capability-based Information Technology Infrastructure Library courses.

Maj. Gen. **Harold W. 'Punch' Moulton** II (ret.) is the former director of operations, US European Command, Patch Barracks, Stuttgart, Germany. The USEUCOM mission is to maintain ready forces to conduct the full range of operations: enhance transatlantic security through support of NATO, promote regional stability, counter terrorism, and advance US interests in the European area of responsibility. Moulton works as Senior Director of Cyberspace Strategies, Integration, and Consulting at Stellar Solutions.

Charles Pak has taught information systems (IS) courses for over 25 years as an IS practitioner and professor. He has managed US federal government data centers for over 20 years, including personnel. He has designed, tested, implemented, and maintained many of these enterprise network sites. These sites are some of the largest in the world and encompass distributed sites across the US as well as international sites. He has managed state-of-the art systems for military and federal government missions for which he was deployed.

James Stavridis has been the dean of The Fletcher School since its founding in 1933. A retired admiral in the US Navy, he led the NATO Alliance in global operations from 2009 to 2013 as Supreme Allied Commander. He also served as Commander of US Southern Command, with responsibility for all military operations in Latin America from 2006–09. Stavridis has published five books and over a hundred articles on innovation, strategic communication and planning, and creating security through international, interagency, and public/private partnerships in this turbulent 21st century.

Constance P. Uthoff is an assistant professor and assistant director of the Master's Degree in Strategic Cyber Operations and Information Management at the George Washington University. She cumulates over ten years of physical and business security experience and has taught courses and seminars on cyber warfare, CND fundamentals, and cyber law and policy. Recently, she co-authored *Project Cyber Dawn*, a cyber analysis of Libya, and she is working on a cyber-intelligence analysis project for the Cyber Security Forum Initiative.

1
Trends in Cyber Operations: An Introduction

Frederic Lemieux

Introduction

In the wake of several historical data breaches in the United States, in early 2015, the White House announced a new series of legislative proposals aimed at securing cyberspace and issued cybersecurity guidance to government agencies and the private sector (The White House 2015). Through this legislative exercise, the federal government wanted to address three priorities: (1) enable cybersecurity information sharing across private organizations and government agencies; (2) modernize law enforcement capabilities to conduct cyber investigations; and (3) establish a nation data breach reporting protocol for businesses that have experienced an intrusion during which personal information has been exposed. Through their implementation, these legislative measures will result in the deployment of both defensive and offensive strategic cyber operations by the government and private industry.

The concept of cyber operation is primarily used in the military field and refers to offensive and defensive activities related to a cyber warfare strategy. According to the US Joint Chief of Staff (2014), cyber operations include, but are not limited to, computer network attack, computer network defense, and computer network exploitation. In reality, cyber operations are conducted across multiple sectors of our society (Lin, Allhof and Abney 2014). For instance, the private-sector finance, telecommunication, and retail industries conduct defensive cyber operations on a daily basis to prevent data breaches or denial of service attacks. Several organizations in the private sector may also perform offensive cyber operations in the form of industrial espionage and competitive intelligence activities (Lin, Allhof and Abney 2014). In the public sector, government agencies including law enforcement, intelligence, and other critical departments conduct both of the aforementioned types of cyber operations by spying on domestic or foreign targets (Schmidt 2014) as well as investigate cyber offenders or provide assistance in protecting critical infrastructure by implementing the Computer Emergency Readiness Team (CERT), for example (Bada, Creese, Goldsmith and Phillips 2014).

In academia, cyber operation is considered a multidisciplinary concept intersecting mostly with the social sciences, behavioral sciences, political sciences, engineering, and law (Shakarian, Shakarian and Ruef 2013). For instance, social scientists study cyber operations from the criminology perspective, conducting cyber criminal investigation and examining illegal activities that occur in cyberspace, such as fraud and identity theft (Stephenson and Gilbert 2013). Behavioral scientists are working to find solutions to network vulnerability by studying human behaviors and developing adaptive cyber operations through biomimetics, for example (Pino, Kott and Shevenell 2014). Political scientists scrutinize current and emerging policy related to cyber operations and examine how state and non-state actors conduct cyber operations and exercise influence on international relations (Erickson and Giacomello 2007). Engineers research and develop new technologies and enhance existing tools that enable the conduct of defensive and offensive cyber operations (Bodeau and Grobart 2011). Lawyers study the evolution of laws related to cyber security and advise lawmakers on new legislation that will regulate cyber operations (Schmitt 2013). Indeed, these academic disciplines interact with each other and shape the way cyber operations are conducted.

Another critical characteristic of the cyber operation concept is its nature, which is both strategic and tactical (Andress and Winterfeld 2011). Tactical cyber operations involve techniques and practices used by information technology professionals to secure or penetrate a computer network. Tactical cyber operations can also be performed by offenders who crack, hack, and breach an information system. Strategic cyber operations build on the approaches that align with the defensive and offensive dimensions. For instance, defensive strategic cyber operations are planned and carried out based on the goals of prevention and deterrence. Offensive strategic operations are usually developed based on more hostile goals. For the purpose of this book, a cyber operation is defined as having a set of comprehensive cyber operational goals that are carefully designed and planned to serve a long-term offensive or defensive purpose. Strategic cyber operations can take the form of policy, strategy, and best practices related to computer network attack, computer network defense, and cyber security incident management.

This introductory chapter is divided into four sections. The first section examines global trends of cyber operations and focuses on current as well as emerging threats in cyberspace. The second section offers an analytical perspective on the intensity of cyber operations and the type of actors evolving in cyberspace. The third section outlines the emerging and most pressing challenges in cyberspace. Finally, the fourth section introduces the structure of the book.

Global trends in cyber operations

Recently, an article in *Time* magazine (Rayman 2014) listed five hotspots in the world for cyber crime and cyber operations: Russia, China, Brazil, Nigeria, and Vietnam. According to the magazine, each hotspot has its particular expertise in terms of criminal capabilities. For instance, Russian cyber criminals are known for being highly skilled in hacking and breaching data systems primarily for profit (mostly for organized crime interests). Conversely, in China, most hackers are not working for organized crime but are operating under the guidance of the government. Chinese hackers are often involved in economic and politic espionage operations. Hackers in Brazil seem to follow the path of their Russian counterparts and have been involved in large-scale money theft and fraud through payment systems as well as by targeting individuals. Cyber criminals from Nigeria are well known for email scams and hacking tactics to extort money from their victims. Finally, the situation in Vietnam presents a hybrid form of what can be found in China and Russia. While a vast number of Vietnamese cyber criminals are involved in data breaches and theft of personal information from Europe and United States, they are also deeply involved in spying operations on neighboring countries and their own citizens for the benefit of the Vietnamese government.

Several countries have experienced an intensification of cyber attacks in recent years. According to the Government Accountability Office (2013), the United States, one of the most targeted countries in the world, has faced a staggering increase of reported attacks on US federal agencies ranging from 5,503 in 2006 to 48,562 in 2012 (see Figure 1.1). Global trends of malicious cyber operations are tracked annually by anti-virus corporations such as McAfee, Symantec, and Kaspersky Lab. Each year, these organizations publish cyber threat assessments and provide statistics related to several types of attacks, targets, and modus operandi. According to Symantec's *Internet Security Threat Assessment Report* (Symantec 2014), 2013 was characterized as the worst year on record

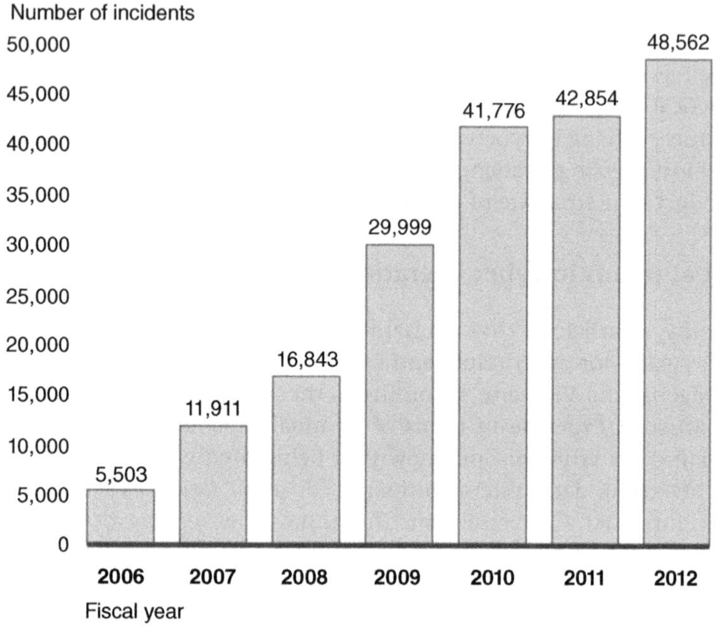

Figure 1.1 Numbers of attacks on US federal agencies between 2006 and 2012
Source: GAO analysis of US-CERT data for fiscal years 2006–2012

for large-scale data breaches. The report also describes several additional important trends. Targeted attacks are on the rise, and the odds of government agencies and manufacturing being targeted is high (the odds are 1/3.1 and 1/3.2 respectively). Mobile capabilities are now plagued by social media scams and malware. According to Symantec's report, cyber criminals have victimized 38 percent of mobile users. Ransomware attacks have increased by 500 percent, and hackers are now moving toward evolved methods called ransomscrypt. Lastly, attackers are now looking at a new field of operation and have started to hack common electronic devices that are part of the 'Internet of Things' (IoT), such as baby monitors, security cameras, and routers.

In its threat prediction for 2014, MacAfee highlights a few more growing trends that posed concerns for governments and industries. The deployment of corporate applications in 'the cloud' will generate new attacks and unsuspected entry points. McAfee estimates that 80 percent of business users are operating applications in the cloud without informing their own corporate IT. Attacks through social media platforms will increase and become more sophisticated using features like location to target victims. According to McAfee, the Pony botnet was responsible for stealing millions of passwords from users on Facebook, Google, Yahoo, and others. The high prevalence of false or fake profiles on social media provides an indication of the capacity of social attackers. Facebook admitted that 50–100 million accounts are duplicates, and a recent survey conducted by Stratecast (2013) indicates that 22 percent of social media users have experienced security issues.

Both private sector and government reports on cyber attacks indicate that attacks motivated by cyber criminals (for profit) and hacktivists are at the top of the list. Also, reports from the anti-virus industry reveal that government agencies, manufacturing, and finance sectors are at the most risk of experiencing attacks. In terms of attack types, defacement, distributed denial of services, SQL injections, and account hijacking were the most frequent malicious attacks between 2012 and 2014. Finally, according to Ponemon Institute (2014) the average cost of a data breach occurring in the United States in 2013 was estimated at $5.4 million. In 2014, the average annual cost of cyber attacks was estimated at $12.7 million, according to a survey of 59 large US firms, indicating a 96 percent cost increase compared to the past five years (Ponemon Institute 2014).

However, while the assessments provided by anti-virus corporations are very detailed and based on millions of attack sensors deployed around the world (up to 157 countries), they don't necessarily expose all malicious cyber operations taking place in cyberspace. For instance, the information leaked by Edward Snowden informed the public about activities conducted by the National Security Agency over several years. None of the anti-virus corporations detected the intrusions committed by the NSA in the US communication system nor the intrusions into foreign government information systems. Despite the fact that the American government can justify these intrusions under national security pretexts, many countries targeted by the NSA's programs admitted that there were real economic and political costs to these spying activities. The lack of reporting and perhaps the selected reporting of malicious cyber operations raise the question of ethics in cyberspace and will be addressed further in this chapter.

Cyber operations: intensity spectrum and actors involved

This section places an emphasis on offensive cyber operations and provides a theoretical approach to categorize the level of intensity of the operations as well as the type of actors that engineer them. Four levels of intensity can be identified, ranging from the least to the most aggressive action against actors. The first level, *passive*, is the least hostile type of cyber operation and can be associated with cyber espionage or reconnaissance activities aimed at gathering information for competitive purposes, for example between state actors or corporations (Hunker 2010a). In this scenario, states and non-state actors will spy or stalk their target in order to collect critical information that can benefit them. In this particular case, the spy or stalker does not want to be discovered, and its activity will exclusively remain stealth to avoid any potential exposure.

The second level, *provocative*, is more hostile than the previous one in the sense that state and non-state actors will use cyberspace to communicate a message or disclose embarrassing information in order to influence or polarize public opinion. On the one hand, individuals such as Julian Assange, Chelsea Manning, and Edward Snowden leaked a tremendous amount of government information with the objective of publicly embarrassing a state on actions they judged unacceptable. On the other hand, violent groups like the Islamic State (ISIS) will use cyberspace to deliver threatening messages or communicate appeals to recruit new members (propaganda). Finally the case of Sony appears to fall under this category due to the leaking of embarrassing emails and information about its employees and artists in order to intimidate the company regarding the non-release of a satirical movie about the assassination of the North Korean leader (blackmailing). In this category, provocative operations use information or messaging against their target in a public manner in order to provoke a reaction.

The third level, *disruptive*, refers to the perpetration of hostile actions to overwhelm and momentarily paralyze a target. Well-known examples of such hostile operations are the stealing of mass personal information, distributed denial of services (DDoS) attacks, and denial of services (DoS) attacks (Hunker 2010a). These disruptive actions generally aim at directly impacting the day-to-day activities of a target by paralyzing information systems, supply chains, and communication channels. These attacks often overwhelm the victim in its capacity to respond and mitigate the consequences of the disruption. For example, before employing traditional warfare operations, Russia is accused of using DDoS against servers in Estonia and Georgia, thereby paralyzing critical systems, such as government websites, the financial sector, and telecommunications. These Russian cyber operations disrupted the ability of Estonia and Georgia to foresee and respond to traditional military aggression by overwhelming the respective governments' major infrastructures, undermining governmental authority prior to the use

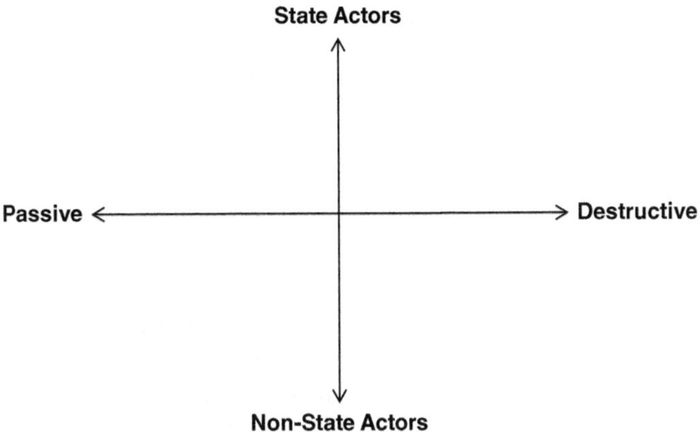

Figure 1.2 Levels of offensive cyber operations and types of actors involved

of kinetic military force (Applegate 2012). In the cases of Home Depot and Target, the stealing of mass credit-card information by hackers led to a prominent slowdown in business, forcing the credit-card issuers to reduce the purchase limit of cardholders and replace all compromised credit cards. In both cases, criminal groups are suspected to have committed the breach and stolen the consumer credit-card information.

Finally, the fourth level, *destructive*, refers to cyber operations that aim at provoking physical destruction of computer systems or any system operating with coded signals over communication channels. These attacks can potentially cause harm to human beings, especially if targeting critical infrastructures (Lin 2010). The most sophisticated example of a destructive cyber operation is the Stuxnet virus, which was used to physically destroy several centrifuges serving to enrich uranium at Iran's Natanz nuclear facility. This destructive cyber operation disclosed how vulnerable critical infrastructure can be if a virus or malware enters a supervisory control and data acquisition system (SCADA), causing large-scale damage.

Cyberspace is composed of a myriad of actors conducting offensive and defensive cyber operations. They can be categorized in two major groups: state and non-state actors (Valeriano and Maness 2014). The level of social organization will differ in each group (see Table 1.1). For instance, non-state actors could be a sole individual who can decide to attack a target because of the challenge it represents, for vengeance, or simply because of greed (fraud). Non-state actors can also be composed of more complex social organizations, such as violent groups or a

Table 1.1 Types of cyber actors according to their level of social organization

	Non-State Actors	State Actors
Low Social Organization	Individual hackers and crackers	Proxy state actors, such as cyber mercenaries
High Social Organization	Corporations, organized crime, terrorist groups	Government agencies and multi-governmental organizations

collective of hacktivists that will attack a target for a moral or political cause. Also, non-state actors can be corporations that decide to conduct offensive operations against competitors or a government agency to steal information critical to the conduct of their business.

The second group, state actors, also varies in its composition. For instance, several states may decide to conduct offensive cyber operations by using a proxy and outsource the entire mission to a non-state actor (company or collective). This type of scenario can be compared to 'guns for hire' or mercenaries, which in this case would be state-sponsored hackers (Schmitt and Vihul 2014). Also, state actors can conduct offensive cyber operations throughout their government agencies that specialize in communication and signal intelligence. For instance, the United States conducts offensive cyber operations through US Cyber Command and the National Security Agency. In the United Kingdom, the Government Communications Headquarters (GCHQ) and Security Service (MI5) are involved in cyber operations. In Canada, the Communication Security Establishment (CSEC) conducts strategic cyber operations in collaboration with the United States and other allied countries. More recently, several states have expressed interest in developing cyber capabilities through supra-national entities such as NATO and the European Union (Herzog 2011; Yost 2010). These discussions are leading to the development and alignment of national doctrines, cyber defense strategies, and offensive capabilities for cyber warfare.

Conceptualizing cyber operations on a spectrum based on intensity provides two analytical advantages. First, it offers a theoretical gradation that categorizes offensive cyber operations taking place in cyberspace, which can be critical in understanding the progression of an adversary vis-à-vis its capability to innovate and conduct sophisticated actions. Second, this spectrum can be intersected with another concept, such as the type of actors operating in cyberspace. The second level of analysis provides a richer investigative design that allows experts to further categorize types of actors according to the intensity of their cyber operations. For instance, only a team of two state actors (US and Israel) has been able to design, deploy, and activate a cyber weapon (Stuxnet) that caused physical destruction, so far the highest intensity on the proposed spectrum in Figure 1.2. However, more research is needed to test the assumption of a correlation between the intensity of cyber operations and the type of actors evolving in cyberspace (including their level of social organization).

Emerging challenges

Over the past decade, we witnessed a growing interest in cyber security problems among Western nations, many of which recently adopted new national cyber security strategies. One of the main challenges following the adoption and implementation of such policies is the common articulation between countries. Differing political and judicial systems limit the coordination of efforts against transnational cyber threats. Most national cyber security strategies are based on a home country's legal principles, which may not apply to other countries, thus restricting the level of cooperation during police investigations or deployment of counter-measures and limiting the scope of incident response (Hunker 2010b; Tikk 2010). This lack of harmonization between countries' national cyber security strategies poses a real challenge that limits the ability of state actors to secure cyberspace against threats propagated on a global scale.

Another important challenge is the availability of and professional development of the cyber security workforce. During the past decade, the demand for cyber security specialists has increase sharply, but the supply of qualified workers remains insufficient to respond to the needs of government agencies and the private sector in the United States and abroad (Evans and Reeder 2010; Libicki, Senty and Pollack 2014). Despite the fact that thousands of positions are still unfilled in 2015, and thousands more will be created in the years ahead, most developed countries have not developed a comprehensive strategy to prepare the cyber security workforce. For the few countries that did launch educational initiatives in cyber security during the past years (such as the United States), recent studies still show a discrepancy between skills developed in training or education programs and industry requirements (National Research Council 2013). Other issues within the workforce emerge as problematic, such as the lack of racial and gender diversification as well as salary discrepancies between the public and private sectors. The development of a cadre of skilled and educated cyber security professionals can also contribute to the effectiveness and integrity of the future workforce.

Ethics and information warfare are other critical challenges with respect to use of cyber weapons, information technology, privacy, and surveillance of communications (Floridi and Taddeo 2014). The case of Edward Snowden and the NSA is revealing about ethical dilemmas in cyberspace (Landau 2014; Lucas 2014). First, it raises the question of government agency accountability in terms of publicly funded spying programs and what kind of sanctions can be imposed if laws are transgressed. For instance, the leaking of national security information by Edward Snowden sparked a debate in which politicians and public opinion were highly divided about the characterization of his actions: traitorous or patriotic (Tavani and Grodzinsky 2014). This debate illustrates a higher moral dilemma between (a) denouncing what was perceived as government misconduct and (b) protecting national security interests. Second, if we admit that espionage between countries is an acceptable practice, how far can a state go to maintain information superiority? Can it compromise the security of its own national industries and allies by inserting a flawed encryption algorithm into a proprietary software product, thereby providing a convenient backdoor into telecommunication systems while at the same time creating a critical vulnerability that could be exploited by adversaries? The NSA is suspected of paying $10 million to RSA, a US computer security company, to implement two faulty encryption algorithms in software that were distributed widely around the world (Menn 2014). These flawed encryptions made them easier to exploit because of substandard requirement for algorithm. Finally, ethics in cyberspace is also challenged by groups of vigilantes and hacktivists who claim pursuit of a moral cause by leaking embarrassing information about governments, corporations, and individuals (Himma 2008). The ethical dilemma underlines the fine balance in exposing injustice or abuse of power by committing a criminal infraction (hacking systems and leaking information).

Another important challenge for state actors is the use of cyber operations as part of their larger warfare strategy (Applegate 2012; Hoffman 2007). For example, Russia attacked Estonia, Georgia, and Ukraine using simultaneously and consecutively deployed cyber weapons and kinetic military action together. This fairly new trend raises the question of the application of international law in cyber warfare operations. Related to this challenge is the principle of proportionality in warfare strategy (Goldsmith 2013; Jensen 2012). Since most offensive cyber operations do not lead to destruction, it is difficult to effectively assess the real consequences of a cyber attack and what is considered a prohibited response (for instance, retaliating by disabling a country's energy network that supplies a large population in a cold winter). Finally, the determination of which actors should take part in cyber warfare operations is still unclear (Lin, Allhoff, and Abney 2014; Schmitt and Vihul 2014). Several governments conduct offensive cyber operations using non-state actors as a screen to mask or defeat the attribution process. Also, in some circumstances, a corporation can decide to respond or retaliate to an attack perpetrated by a state actor resulting in the fueling of a conflict – with or without knowledge of that existing, ongoing conflict.

Another important challenge that should be highlighted is the use of cyber operations by terrorist groups (Gilmour 2014; Singh and Krupakar 2014). For several years now, groups such as Al Qaeda and Al Shabaab have used the Internet to distribute their propaganda through videos and websites (Stenersen 2008). They also use chatrooms to recruit and radicalize what are now known as homegrown terrorists (Awan 2007). More recently, the group Islamic State of Iraq and Levant (ISIL) has started using more sophisticated and aggressive tactics in cyberspace. For instance, the quality of recruitment video clips released on the web has reached a high level in terms of production and photography (Neer and O'Toole 2014). Also, the staging of the execution of several foreign workers and journalists recorded on camera and released with political messages targeting Western countries is now a common modus operandi of ISIL. The ISIL-affiliated group called CyberCaliphate was able to target a Maryland news station and deface its website and Twitter account to broadcast its propaganda by posting the message 'I Love You ISIS.' The CyberCaliphate was also able to deface the US Central Command Twitter account and post the same message (Cooper 2015). This social media defacing is probably one step in a broader strategy that will certainly lead to an increase of the intensity of its cyber operations.

Book structure

This book provides original perspectives from scholars and high-profile practitioners working in the cyber domain. Theoretical and practical approaches are proposed to understand the evolution of cyber operations and provide multidisciplinary examination of trends around the globe. This book aims at providing an understanding of current changes and issues in cyber operations to inform policymakers, homeland security leaders, military operations, private-sector industry professionals, and graduate students.

This book offers several distinctive features. First of all, cyber operation is a fairly new topic and professional field that encompasses several disciplines associated with technical and non-technical operations conducted in cyberspace. By placing an emphasis on strategic cyber operations, this book's primary focus is on non-technical aspects of cyber operations, such as policy, strategy, and best practices. Secondly, the book is a collection of chapters written by well-established scholars and recognized leaders in the industry, providing a unique amalgamation of policy, strategy, and practice developed by credible experts in the domain. Thirdly, the book embraces an international perspective by addressing issues related to cyber operations from different regions of the world. Finally, the book presents cyber operations not only in a military (warfare) context but also in the civilian environment (private industry). The book is divided into four sections: (1) conflicts in cyberspace; (2) geopolitics of conflicts in cyberspace; (3) cyber defense strategies and practices; and (4) cyber intelligence and information security.

The first section provides an overview of current and emerging trends related to conflicts in cyberspace. In this section, Scott Applegate (Chapter 2) provides an overview of cyber conflict and its application to state and non-state actors (countering terrorists and criminal organizations). By using case studies, the chapter explores how warfare strategies and tactics have been applied to cyberspace. The readers will also be introduced to innovative warfare concepts and doctrines as they relate to cyber security. The chapter scrutinizes existing international and domestic issues related to cyber aggression and examines in-depth case studies. This chapter provides a rich context for discussion about emerging coordination strategies and innovative decision-making. As the second contribution in this section, Andrew Colarik and Lech Janczewski (Chapter 3) examine the theoretical underpinning of current cyber warfare research, what has been learned so far about its application, and some of the emerging themes to be considered. The authors also postulate the development of a (national) cyber warfare doctrine (CWD). An endeavor of this scale requires lots of considerations and preparation for its development if it is to be cooperatively embraced. This chapter considers why information technology systems and their supporting infrastructures should be considered legitimate military targets in conflicts and offers several situations that support this supposition. In addition, it identifies the various forms of doctrine that will become the basis for developing a CWD, discusses a CWD's possible components, and proposes a national collaborative and discussion framework for obtaining a nation's stakeholder buy-in for such an endeavor. Finally, Jack Goldsmith (Chapter 4) explains these vulnerabilities, sketches how they affect the laws of war, and conjectures that international norms to regulate and temper attacks on these vulnerabilities are unlikely to develop. The chapter also provides an assessment of the ways that the rise of cyber exploitation and cyber attacks challenge prevailing conceptions of the laws of war.

In the section titled 'Geopolitics of Conflict in Cyberspace,' Roland Heickero (Chapter 5) explores Russian views on information warfare (IW) and strategic information operations. The chapter provides an overview of Russia's information-warfare capabilities, ambitions and practices. The author describes, analyzes, and discusses information operation strategy and doctrine. Russian organizations responsible for information warfare are examined, as well as a brief comparison between Russia's, the United States', and China's strategic information operations. Using the historical examples of Estonia and Georgia, the chapter examines how Russia is able to integrate information operations into more conventional military warfare practices. Lastly, the chapter addresses current and emerging threats as well as malicious activities that originate from Russia, such as cyber criminality and hacktivism. In the following chapter, Jyh-An Lee (Chapter 6) illustrates the dangers that China-based threats pose to the governments and enterprises worldwide, especially for the United States. The author analyzes several proposals that aim at providing solutions that minimize the international cyber security problem, and the author discusses their possible implementation. The chapter also focuses on the positions of the United States and China, as the two powers are likely to have the most complicated interests at stake and the greatest influence on the issue of security in cyberspace. Finally, Jeffrey Bardin (Chapter 7) contends that jihadist and belligerent nation-states located in the Middle East are rapidly learning how to use asymmetric warfare through the Internet and how to more successfully conduct information operations against Western targets. The chapter explores passive information operations, such as recruitment, martyrdom, anti-West propaganda, and awareness campaigns about Western foreign policies. The chapter also places an emphasis on more aggressive information operations that target Western systems and infrastructures. For example, the author discusses the emergence of cyber military divisions and cyber warfare capabilities from Iran and Syria.

In the third section, titled 'Cyber Defense Strategies and Practices,' Harold Moulton, James Stavridis, and Constance Uthoff (Chapter 8) explore the idea of incepting a national strategy for cyberspace throughout the implementation of a federation that can comprehensively advocate, organize, facilitate, and protect national capabilities in cyberspace. By using examples such as the FAA and NASA, the authors argue that such a 'cyber federation' can singularly focus on cyber operations across all its many facets: technical, risks, challenges, and opportunities. The authors contend that a new organization guided by a clear policy on cyberspace can leverage private and public expertise with a national mandate to enhance the state's competitiveness and security in the cyber arena. In the next chapter, Frederic Lemieux (Chapter 9) proposes an approach to create a trustworthy cyber defense information-exchange platform that will enhance cooperation within and across the critical infrastructure sectors to better share cyber security information. This data-exchange platform pursues three main goals: (1) developing a data-exchange platform housing cyber security data in a trustworthy, cooperative environment; (2) resolving existing limitations related to law and policy, security, and automation of exchanges in cyber security through inception of advanced technology that enhances information-sharing confidence; and (3) creating an incentivizing data-exchange process by providing advanced analytical capabilities and information access that will support the operational and strategic decision-making of participants. Finally, Wayne Harropp and Ashley Matteson (Chapter 10) present cyber resilience as a key strand of national security. This chapter establishes the importance of critical national infrastructure protection and the growing vicarious nature of remote, well-planned, and well-executed cyber attacks on critical infrastructures. The authors present examples of well-known historical cyber attacks and explore the emergence of the 'Internet of things' as one of the cyber vulnerability issues yet to be addressed. The chapter identifies key steps being undertaken by those responsible for detecting, deterring, and disrupting cyber attacks on critical national infrastructure in the United Kingdom and the United States of America.

In the fourth section, titled 'Cyber Intelligence and Information Security,' Charles Pak (Chapter 11) introduces the readers to information systems' vulnerabilities and cyber crime, such as theft of sensitive information, compromise of computer networks, identity theft, cyber attacks, and intrusion-detection systems as well as a hacker's modus operandi. The chapter also addresses different challenges related to national critical infrastructure. Zal Azmi's work (Chapter 12) explores the importance of 'Big Data' in solving complex and emerging problems that rely on combining a massive amount of data from different data sources in real time while maintaining the security of the data sets. More precisely, Azmi examines how the volume and variety of information (text, audio, video, tweets, and the like), combined with the speed with which it is being generated, have challenged security experts and security technologies to maintain the confidentiality, integrity, and availability of information. Finally, Constance Uthoff (Chapter 13) examines definitions and models of cyber intelligence use across industry, the government, and military through a review of current dialogue, reports, and processes related to cyber intelligence tradecraft, analysis, and application. By understanding the application of various cyber intelligence frameworks, decision makers can more fully select strategies that meet their organizational needs and anticipate future threats with greater certainty. These insights can also lead to the development of more effective approaches and the design of supporting technology.

Section I

Conflicts in CyberSpace

2
Cyber Conflict: Disruption and Exploitation in the Digital Age

Scott Applegate

Introduction

In August of 1986, a former astronomer-turned-systems-administrator, on his second day on the job, attempted to determine what was causing a 75-cent discrepancy in a UNIX accounting system at the Lawrence Berkeley National Laboratory. Over the course of the next ten months, Clifford Stoll and his coworkers would trace the anomaly to a hacker in Germany who was using computers to steal information from the United States and sell it to the Soviet Union's KGB (Stoll 2005). This incident represents one of the earliest examples of cyber espionage and pioneered many of the tactics, techniques, and procedures used in cyber incident response actions today. More importantly, this event demonstrated the ability of nation-states or their proxies to leverage emerging network technologies to gather intelligence and to potentially disrupt the services and systems of competitor states. It was one of the groundbreaking events in a phenomenon we will collectively call 'cyber conflict.'

In the two and a half decades since Mr. Stoll's efforts, network technology and the Internet have grown exponentially and have become a pervasive part of everyday life. This growth has come at the cost of increased vulnerability and risk, as nation-states have become progressively more reliant on complex network technology for everything from government and military operations to medical and financial transactions. Complexity breeds vulnerability in the world of information technology, and while these new technologies have empowered users to access information and operate in ways previously not possible, they have also exposed users to new forms of attacks and criminal activity. Individual hackers have exploited vulnerabilities in information technology in order to conduct cyber attacks ranging from defacing web pages to industrial sabotage. Bands of hackers have joined together to form politically or ideologically motivated groups such as Anonymous and Lulzsec, leveraging the power of the Internet to make their voices heard, shape political dialogs, and attack individuals, organizations, or even nation-states with which they disagree. Terrorist groups are employing the Internet to recruit, train, motivate, and synchronize their followers. Nation-states, recognizing the power of this domain, have begun building cyber-warfare programs and training their own cyberspace warriors to identify and exploit weaknesses in the military, government, and commercial networks of their competitors (Alexander 2007).

All of this activity can collectively be described as cyber conflict and represents an evolving phenomenon of military- and politically-motivated actions, leveraging cyberspace as a medium to attack and exploit competitors' information systems and resources. 'Cyber conflict' can be defined as activities that occur in cyberspace that involve armed or unarmed hostilities or similar adversarial actions between nation-states, or between nation-states and organized non-state groups. This broad definition encompasses cyber warfare, cyber espionage, cyber terrorism, and some forms of hacktivism. The first step to understanding the concept of cyber conflict is understanding the domain in which it takes place and how that domain has evolved from the fledgling networks of the 1970s and 1980s into a global commons and full-fledged war-fighting domain under current United States military doctrine.

Defining the domain

The term 'cyberspace' first started to appear in print in science-fiction stories of the early 1980s. William Gibson was not the first author to use this term in print, but he is generally credited with its invention and association with online networking. Gibson defined cyberspace as 'a consensual hallucination experienced daily by billions of legitimate operators, in every nation, by children being taught mathematical concepts... A graphic representation of data abstracted from the banks of every computer in the human system. Unthinkable complexity. Lines of light ranged in the non-space of the mind, clusters and constellations of data' (Gibson 1984).

While Gibson's fictional vision of cyberspace may not exactly match the reality of cyberspace today, it does reach billions of users in every nation and has become almost pervasive in nature. Years later, in an interview, Gibson noted that when he coined the term 'cyberspace,' 'it seemed like an effective buzzword. It seemed evocative and essentially

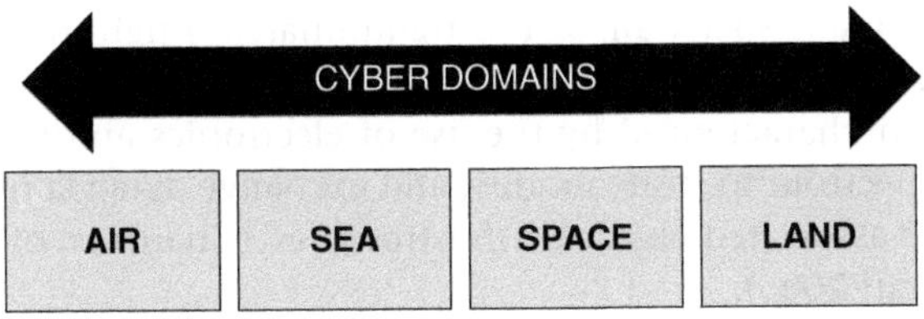

Figure 2.1 Cyber domains

meaningless. It was suggestive of something, but had no real semantic meaning' (Neale 2000). This description also remains mostly true today. Ask three different people to describe cyberspace and you will get three different definitions. As cyberspace has evolved from the fledgling networks of the 1970s and 1980s to the global internets of the 1990s and beyond, its importance to government and military operations has also grown tremendously. Military operations in cyberspace initially focused on data and information. In recent years, this focus has evolved into a rapid militarization of cyberspace with multiple nation-states initiating programs to develop their war-fighting and intelligence-gathering capabilities in this domain.

Information and intelligence has always been an important aspect of military operations. However, the formal doctrinal development of cyberspace as a war-fighting domain can be traced back to the 1990s, when military strategists began to understand the importance of information technology in modern warfare. American military dominance during the 1991 Gulf War demonstrated the power that network-centric warfare (NCW) provided commanders on the battlefield. Even in its nascent state in the 1990s, NCW allowed commanders of highly mobile and geographically dispersed forces to share and maintain situational awareness, synchronize movement, and consistently outmaneuver enemy combatants. Other nation-states, such as China and Russia, took note of this victory and immediately began exploring these emerging concepts in order to incorporate them into their own operations.

In September of 2001, the United States Department of Defense (DOD) published *Joint Publication 3-0: Operations*, with the landmark inclusion of 'information' as a fifth war-fighting domain on equal footing with air, land, sea, and space (Kelley 2008). In 2006, the DOD published the since-declassified *National Military Strategy for Cyberspace Operations* (NMS-CO), which redefined the information domain as cyberspace and noted that 'treating cyberspace as a domain established a foundation to understand and define its place in military operations' (Chairman of the Joint Chiefs of Staff 2006). As depicted in Figure 2.1, this new domain spans all of the traditional war-fighting domains, interacting with and affecting each. The NMS-CO defined cyberspace as 'a domain characterized by the use of electronics and the electromagnetic spectrum to store, modify, and exchange data via networked systems and associated physical infrastructures' (Chairman of the Joint Chiefs of Staff 2006).

While this was just considered a working definition, it failed to adequately address key elements of this domain and has since been updated. Under current US military doctrine, cyberspace is defined as 'a global domain within the information environment consisting of the interdependent network of information technology infrastructures and resident data, including the Internet, telecommunications networks, computer systems, and embedded processors and controllers' (Chairman of the Joint Chiefs of Staff 2014). Even this latest definition leaves out a critical aspect of cyberspace – the human being. Cyberspace is more than just information and the technology connecting it, it is also comprised of the people who interact with it and each other. Cyberspace has been described as having three aspects or layers: a physical layer, a logical layer, and a cognitive or social layer.

The *physical* layer of cyberspace consists of the information systems, servers, hardware, wires, cables, routers, and even the radio frequencies that make up the network architectures of the Internet and other interconnected and closed networks (United States Army Training and Doctrine Command 2010). While cyberspace is often viewed as a boundless global commons, it is important to remember that every device that makes up this domain physically exists and, more importantly, the vast majority of these devices are physically located within the geographic boundaries of a sovereign nation-state. This can have significant implications in terms of legal authorities to operate in this domain and in terms of jurisdiction in response to actions taken in this domain as well.

The *logical* layer of cyberspace consists of the connections between nodes where nodes are essentially any device or service represented by an Internet protocol (IP) address. The logical layer of cyberspace provides a set of capabilities that are intentionally divorced to a great extent from the details of the technology that underpins it (Clark 2010). For instance, an individual using a Mac computer to send an e-mail message to a friend using a Microsoft Windows computer does not need to worry about the complexity of making these two different operating systems communicate with each other. That complexity is hidden from both users. The logical layer provides a layer of abstraction that allows dissimilar devices to communicate with each other in a manner that is transparent to the users. It is this layer that also gives rise to many of the vulnerabilities that are exploited in the various systems that make up the Internet.

The *cognitive* layer of cyberspace consists of the information and the human beings that interact with it. This layer is also sometimes called the social layer. As noted above, people are a critical aspect of cyberspace. 'All operations in cyberspace begin with a human being' (Cyber Intelligence Task Force 2013). People are not just passive users but define and shape this domain through their interaction with it (Clark 2010). The cognitive layer also includes the information stored, transmitted, and processed on the Internet in all its myriad forms, from raw data to highly processed information.

There is a tendency to focus on the technical aspects of cyberspace rather than holistically embracing all aspects of this domain. Administrators are often much more comfortable addressing highly technical vulnerabilities than they are in dealing with the personnel who use these systems. It is critical, however, to keep in mind the relationship between the physical, logical, and social layers of cyberspace when conducting operations in this domain. To clarify the relationship between these three layers of cyberspace, we can look at a single web server as an example.

A single server may host multiple websites from the same physical location. In this example, the physical layer is the physical location of the server and its hardware. The ability to host multiple websites illustrates the logical layer wherein a user may enter one uniform resource locator (URL) address to retrieve a particular website from this physical server, while entering a different URL may retrieve a different website from this same physical server. The user may never realize these websites are hosted from the same physical location, as the logical layer separates this technical detail from the user. Finally, the information stored on this server, the users accessing it, and even the interaction of these individuals with this server, comprises the cognitive layer.

Understanding these three layers of cyberspace is important as they each represent different vectors that an attacker can use to exploit or disrupt capabilities that are reliant on or make use of cyberspace. An attacker wishing to disrupt websites on the previously discussed server could physically damage or destroy the hardware. He or she could disrupt the site at the logical layer by exploiting vulnerabilities in the software or by redirecting traffic away from the websites. Lastly, an attacker could focus on the cognitive layer, attempting to deceive the website's users through social engineering (phishing or deceptive e-mail, for example) or other techniques.

Defining cyber conflict

Having defined the domain in which cyber conflict takes place, we must now take a closer look and define what actually constitutes this collective phenomenon. As noted earlier, cyber conflict can be defined as deliberate activities that occur in or through cyberspace that involve armed or unarmed hostilities, or similar adversarial actions, between nation-states, or between nation-states and organized non-state groups. In some limited cases, these activities may take place exclusively between non-state groups but may rise to a level that could impact the political, economic, or diplomatic interests of a nation-state. 'Cyber conflict' is an overarching term that encompasses cyber warfare, cyber espionage, cyber terrorism, and some forms of hacktivism and patriotic hacking. To understand cyber conflict, we must understand each of the activities that comprise it.

Cyber conflict activities could also be broadly divided further into two subcategories: disruption and exploitation. Disruption activities encompass cyber warfare, cyber terrorism, and some aspects of hacktivism and can be defined as cyber activities intended to deny, damage, disrupt, or destroy information resources and their underlying architectures or other connected technologies. Exploitation activities encompass cyber espionage and some aspects of hacktivism and are focused on exploiting information technology to steal various forms of information and data. Disruptive activities generally receive the most public attention, especially from the popular media, so we will begin our exploration with this category of cyber conflict activity.

Hacktivism

'Hacktivism' is a controversial term used to describe activities carried out by loosely-affiliated online activists, ranging from harmless online sit-ins to malicious or destructive actions that undermine the security of the Internet as a technical, economic, and political platform (Krapp 2005). More innocuous definitions would characterize hacktivism as expressive politics, free speech, human rights, information ethics, information liberation, or electronic direct action toward social change by combining programming skills with critical thinking (Krapp 2005). However, hacktivism more frequently involves actions such as use of malicious software, defacement of websites, construction of false-mirror sites, and diversion of web traffic, or flooding of servers with distributed denial of service (DDoS) attacks (Krapp 2005).

Hacktivist groups like Anonymous have actually created easy-to-use tools, such as the Low Orbit Ion Cannon, to enable relative novices to participate in these types of cyber attacks. Hacktivists are also known for penetrating corporate or government systems, exfiltrating data, and then exposing that data through public releases of embarrassing, proprietary, or classified information in the furtherance of their varied ideological agendas.

Hacktivism falls under the banner of cyber conflict because, more often than not, the targets of hacktivists are nation-states and their respective governments. The earliest known example of hacktivism dates back to October 1989, when the US Department of Energy, High Energy Physics Network, and the National Aeronautics and Space Administration (NASA) were penetrated by Australian anti-nuclear hacktivists who unleashed the Worms Against Nuclear Killers (WANK) worm, which altered login screens (Dreyfus and Assange 2012). The WANK worm displayed random preprogrammed messages to users of infected systems, and in some cases tricked users into believing their files were being deleted, but did no real damage to infected systems (Assange 2006).

While this early example was fairly innocuous, the capabilities of hacktivist groups have grown tremendously over the intervening decades. In recent years, hacktivist groups have been very influential in political movements such as the Arab Spring, where groups like Anonymous conducted cyber attacks and other activities in support of the popular uprisings in Egypt, Libya, Syria, and Tunisia (Karatzogianni 2013). The power of hacktivists is such that there is evidence that nation-states have begun to employ, or at least tacitly encourage, hacktivists as proxy actors against other states. Many of the actions undertaken by hacktivist groups have been branded as cyber terrorism by popular media outlets and the victims of these activities. However, this is not really an accurate use of this term, as terrorism is distinctly different from hacktivism, and it is important to differentiate between these two activities.

Cyber terrorism

Cyber terrorism is the use of the information systems to conduct or threaten to conduct violent criminal acts in order to induce a state of terror in the general public, in the furtherance of a political, ideological, or religious agenda. Such acts either deliberately target or disregard the safety of neutral parties or non-combatants and are generally intended to have psychological repercussions beyond the immediate target. Thus far, there have been no publicly known events that would rise to the level of cyber terrorism. Still, terrorist groups have made ample use of cyberspace for messaging, propaganda, communications, and training and recruitment of new members. In recent years, groups like Al Qaeda and the Islamic State in Syria and Iraq (ISIL) have leveraged the Internet to spread videos of attacks and the graphic beheadings of hostages.

While there have not been direct terrorist actions conducted in or through cyberspace, the convergence of information technology, along with the emergence of potential cyber-kinetic effects generated solely through the exploitation of information technology, certainly opens up the possibility of such attacks by terrorist groups in the future (Applegate 2013). Given this, the threat posed by cyber terrorism, while unrealized at this point in time, cannot be dismissed out of hand. Additionally, many nation-states are expending considerable resources to deny terrorist groups access to Internet resources and to disrupt their current activities in cyberspace.

Cyber warfare

The terms 'cyber warfare' and 'cyber war' have been consistently overused in cyber conflict. Cyber espionage, cyber attacks by hacktivists, and even cyber crime have all been defined as cyber war by the victims of these attacks, by the popular media, and even by government officials and politicians. News sites described the cyber attacks against the country of Estonia in 2007 as the 'first war in cyberspace' even through the vast majority of those participating in these attacks appear to be civilian hacktivists and patriotic hackers (Mite 2007). Given the gravity of the concept of war, it is very important that this concept in cyberspace be carefully and narrowly defined. Cyber warfare is the use of armed attacks in or through cyberspace as an extension of one nation-state's politics to impose its political will onto another nation-state. In almost every case seen thus far, the activity that is occurring in cyberspace does not meet the criteria to be called war. There are two questions we can ask to determine if an action could potentially be considered cyber warfare:

1 Are multiple nation-states involved?
2 Does the action constitute an 'armed attack' or a 'use of force' under international humanitarian law?

If the answer to either of these questions is 'no,' the action in question is probably not cyber warfare. War is a condition that exists between nation-states. Therefore, if only one state is involved in hostile actions, with non-state actors for example, it cannot really be considered war in the technical sense. Civil wars would clearly be an exception to this rule, but, in general, war is a political state that exists between two nation-states. In terms of international law, the United Nations Charter states that 'all members shall refrain in their international relations from the threat or *use of force* against the territorial integrity or political independence of any state, or in any other manner inconsistent with the purposes of the United Nations' (Schmitt 1999).

Although there is an ongoing debate as to what might actually constitute a 'use of force' in cyberspace, it is by and large a safe assumption that unless an action results in physical damage, death, or injury, or threatens the political independence or territorial integrity of a nation-state, it is not considered a use of force. This, then, leads to the question, have we ever seen any actions in cyberspace that could be construed as cyber warfare? The answer is yes. There are at least two well-known cases of hostile actions in cyberspace that would meet the above criteria: Operation Orchard and Stuxnet.

On 6 September 2007, Israel conducted a bombing raid against a suspected nuclear reactor site in its neighboring state of Syria. Although details of this raid remain mostly classified, what is known is that through the use of electronic warfare and a precision cyber attack, Israel managed to disable Syria's air defense system, one of the most capable in the world at the time, leaving it blind and unable to detect an entire squadron of Israeli F-15I and F-16I aircraft as they entered Syrian airspace, conducted their raid, and then departed (Rid 2012). This combined operation was known as Operation Orchard and demonstrated a textbook use of a cyber attack in conjunction with kinetic operations.

In June of 2010, a cyber weapon known as Stuxnet was discovered by security researchers. This cyber weapon was active between November of 2005 and June of 2012 and appears to have been designed to specifically target and physically damage centrifuges at the uranium enrichment facility at Natanz, Iran (Langner 2011; Symantec Corporation 2014). Although no nation-state has claimed responsibility for this attack, it is widely speculated to have been the work of either the United States, Israel, or a combination of both of these countries as part of a covert cyber warfare program code-named 'Olympic Games' (Sanger 2012). The cyber attacks associated with Operation Orchard and Stuxnet are the only two examples available of cyber warfare. However, given the amount of intellectual and financial capital many nation-states are now investing in their cyber warfare programs, it is very likely that this type of activity will grow substantially in the near future. Currently, China, France, Germany, India, Iran, Israel, North and South Korea, Russia, the

United Kingdom, and the United States are all known to have active cyber warfare programs.

Cyber espionage

There has been a tendency during the last 25 years to over-hype the probability and consequences of disruptive cyber attacks while ignoring the impact of intrusive cyber attacks or exploitation (Healey 2013). The invention and spread of information technology has acted as a tremendous enabler and force multiplier to the intelligence operations of nation-states. These technologies have geometrically increased the threat and impact of espionage on the well-being of nation-states while, at the same time, they have lowered both the threshold for entry and the potential repercussions of such actions to those states which engage in cyber espionage.

Over the last three decades, computer network exploitation (CNE) or cyber espionage has grown tremendously in scope and scale. What started as the acts of individual hackers attempting an exfiltration of data from one nation to sell to another for profit has grown into sophisticated and complex cyber espionage programs conducting CNE on an almost industrial scale. In 2013, estimates of damage due to cyber espionage ranged from $20 billion to more than $120 billion for the United States alone (McAfee Corporation and Center for Stategic and International Studies 2013).

It is nearly impossible to get an accurate estimate of the impact caused by cyber espionage due to the reticence of both private corporations and nation-states to share information related to compromises, along with the difficulty in ascertaining the value of the information that has been stolen. Beyond financial losses, the compromise of military, diplomatic, and economic information can also damage a nation's competitive capabilities in the international community. In 2012, General Keith Alexander, then head of both the National Security Agency and the United States Cyber Command, noted that the loss of industrial information and intellectual property through cyber espionage constitutes the 'greatest transfer of wealth in history' (Segal 2012). Evidence suggests that the most prolific examples of large-scale cyber espionage programs can be found in Russia and China.

While Russian cyber espionage efforts do not make headlines as often as the Chinese ones do, recent discoveries by Crowdstrike, Kaspersky Lab, and BAE Systems highlight the sophisticated capabilities the Russian Federation is able to field in terms of advanced cyber espionage toolsets (Finkle 2014; Kaspersky Lab's Global Research and Analysis

Team 2014; Paganini 2014). The Russians have long been suspected of using cyber espionage as highlighted by both the Moonlight Maze and Buckshot Yankee incidents. However, the recent discovery of the Snake cyber espionage campaign is the first time that the Russian Federation has been tied to cyber espionage that appears to be focused on economic rather than political gains (Finkle 2014).

The Snake cyber espionage campaign, which includes the Epic, Turla, and Uroburos toolsets, is a sophisticated, long-term, persistent effort employing multiple vectors and at least two zero-day attacks in order to compromise targeted networks and systems. Snake leverages spear phishing, social engineering, and watering-hole attacks and appears to be focused on the energy and technology sectors (Kaspersky Lab's Global Research and Analysis Team 2014). Snake is a classic example of a state-sponsored cyber espionage campaign. It is a long-term effort with the earliest indicators of compromise dating back to 2005 and attacks still active in 2014. Systems are compromised through multiple vectors and, once compromised, attackers attempt to quietly establish a persistent foothold to allow for remote access, lateral movement, and exfiltration of data. Loss of data and damages for this type of attack are almost impossible to calculate given that some systems have been compromised for upwards of eight years. The Snake campaign appears to be very similar in design to the Chinese efforts, which have become known as 'advanced persistent threat' (APT).

China has become well known for its cyber espionage campaigns, although it continues to deny involvement in these attacks. China is suspected of carrying out a number of large-scale cyber espionage campaigns, including GhostNET, ShadyRAT, Titan Rain, Operation Aurora, and Byzantine Hades. While the details of these operations remain clouded in secrecy, a recent investigation by the Mandiant Corporation revealed evidence directly tying a Chinese military unit to cyber espionage attacks against 141 separate organizations across 20 industries focused almost exclusively in English-speaking countries around the world (Mandiant Corporation 2013).

The Mandiant Report, as it is popularly known, focused on a group it refers to as APT1, one of the more than 20 advanced persistent threat groups it has identified conducting cyber espionage activities from China. This report presented evidence linking these activities to a specific unit of the People's Liberation Army, Unit 61398, located in the Pudong New Area of Shanghai, and concluded that APT1 had conducted a cyber espionage campaign against a broad range of victims since at least 2006. The Mandiant Report is the first publicly available evidence directly linking the Chinese government to these activities. As a result of this, the US Department of Justice indicted five Chinese military officers on charges of computer hacking, economic espionage, and other offenses targeting six American organizations in the nuclear power, metals, and solar industries (Ackerman and Kaiman 2014). This was the first instance where criminal charges had been filed against known state actors for hacking and marks a turning point in US policy regarding its handling of cyber espionage (The Federal Bureau of Investigation 2014).

For its part, China continues to deny involvement in these events, calling the allegations preposterous and accusing the United States of employing a double standard given the recent revelations of activist Edward Snowden (Kaiman 2014). In a second incident in 2014, a Chinese executive was arrested in Canada and charged with hacking the computer systems of US companies with defense contracts. The suspect allegedly targeted information on US fighter jet and other military programs (Grossman and Yadron 2014). These indictments appear to be an escalation in US response protocol, which previously focused on using diplomatic protests along with *naming and shaming* to curb China's alleged use of industrial espionage.

States, proxies, and the rogue elements of cyberspace

Cyberspace is defined as much by the individuals and groups that utilize it as it is by its underlying architecture. In regard to cyber conflict, users can generally be categorized as either state or non-state actors. The category into which they fall has implications for their legal status under domestic and international laws and how states may respond to actions by these actors.

State actors

State actors can be defined as those individuals, groups, or organizations who act directly under the authority of their respective governments, whether they do so in an overt or covert manner. Examples of state actors include government and military organizations, as well as law enforcement and intelligence agencies. These organizations are directly governed by the legal regimes, policies, and regulations of their respective governments and operate under the specific guidance or orders of responsible and accountable officials. Should the actions of state actors rise to the level of use of force in cyberspace, it is likely the actions of these actors would fall under the purview of the Law of Armed Conflict, and they could be subject to diplomatic, military, or legal repercussions as a result. In short, should a state conduct an armed cyber attack against another state, its actions could allow the victim state to invoke the self-defense clause under Article 51 of the United Nations Charter. Examples of state actors' participation in cyber conflict would include most of the previously discussed cyber espionage campaigns by Russia and China, the Stuxnet attack, and Operation Orchard conducted by Israel.

Non-state actors

Non-state actors are individuals, groups, organizations, or even private corporations who engage in cyber conflict activities in cyberspace without the direct authority of their respective governments. Non-state actors, hacktivist groups, and patriotic hackers often participate in online hacking battles based on political, ideological, or patriotic considerations. While these activities would generally have to be conducted against a nation-state to be considered cyber conflict, there are instances in which these groups' activities against each could raise to such a level as to impact the interests of a nation-state. In these instances, this type of activity would fall under the banner of cyber conflict, as the actions of these groups could potentially draw their respective host states into hostilities.

Private corporations and other commercial entities have also played a major role in cyber conflict. Unlike conflict in the other war-fighting domains, non-state actors, especially private and commercial organizations, often play a decisive role in cyber defense (Healey 2013). If we look at the attack on Estonia as an example, the vast majority of the work in both defending Estonia's networks and restoring its networks back to normal operations in the aftermath of the 2007 attack fell mostly on the private sector. Some private corporations have also begun to actively fight back when attacked. When Google discovered it had been the victim of a suspected Chinese cyber espionage campaign, it secretly began its own counteroffensive to trace the attacks back to mainland China and gather evidence against the People's Republic of China as the source of this campaign (Sanger and Markoff 2010). The concept of private corporations, as well as other non-state actors, conducting their own cyber attacks against nation-states has startling implication for the future of this domain.

While the participation of non-state actors in conflict and hostilities between nation-states is not new, cyberspace and network technologies have enabled non-state actors to engage in these activities at a level previously unseen in the other war-fighting domains and to do so without the risk of physical injury or death. Cyberspace removes the geographic constraint associated with most forms of conflict. For example, in the Iraq War, there was a great deal of participation by terrorist and extremist groups within the geographic boundaries of Iraq, and these actors were able to drive up the levels of violence executed against coalition forces.

The same opportunity for participation was not available to individuals in the US or other nation-states to participate in these hostilities due to the geographic isolation of the battlefield. Non-state actors in these states would have to physically travel to Iraq in order to engage in these hostilities. The same constraint does not hold true in cyber conflict. If we look at more recent conflicts such as Georgia or Ukraine, there are constant cyber attacks conducted by non-state actors on both sides of these conflicts and from third parties scattered across the globe.

A major concern with the participation of non-state actors is that they lack the legal, moral, and ethical constraints generally imposed on state-controlled entities by their respective governments. This lack of constraint could lead to the escalation of state-on-state conflicts and could inadvertently drive hostilities toward what Clausewitz would deem to be *total war* in the cyberspace domain (Clausewitz 1976). These activities could also act as drivers to spur or escalate kinetic conflicts between states. This threat is especially troubling in regional, long-standing conflicts where there is a great deal of nationalism and patriotic, ethnic, or religious zeal. The historical tension between Pakistan and India is a perfect example of this dynamic and one we will explore at length to illustrate this concern.

Non-state groups in India and Pakistan are of particular concern, especially given the volatility of the relationship between these two states, the immaturity of their respective information infrastructures, and the fact that both states possess nuclear weapons. There have been four wars and numerous cross-border skirmishes between India and Pakistan since the creation of these two states in 1947. In addition to traditional conflicts, there have also been numerous terrorist attacks, the most famous of which occurred in Mumbai in 2008 when Pakistani terrorists killed more than 150 people in a series of coordinated shootings and bombings (Schifrin 2009). Inflamed passions and nationalism have also spilled over into cyberspace with cyber militias and hacktivist groups conducting attacks on both sides.

There are a number of prominent hacker groups operating in both India and Pakistan conducting attacks against both one another and their opposing governments. These groups include the Indian Cyber Army (ICA), Indian Cyber Force, Pakistan Cyber Army (PCA), Pakistan

Cyber Force, and the Bangladeshi Cyber Army. The PCA has conducted a number of high-profile cyber attacks in recent years, including hacking into India's Central Bureau of Investigation and India's state-owned telecommunications company, Bharat Sanchar Nigam Limited (Almeida 2012; Sagar 2010). The Indian Cyber Army has conducted similar attacks against Pakistani government websites, including attacks on 36 sites in 2010 and what was described as breaches of thousands of websites (including the Pakistani Department of Education) in 2012 (Kurian 2012; The Express Tribune 2010). The later series of attacks, which the ICA called a 'cyber gangbang,' was specifically described as retaliation for attacks previously conducted by the PCA (Kurian 2012). It is this penchant for retaliation among non-state actors that could prove dangerous and lead to escalation of conflicts between nation-states.

As indicated above, each time one of these non-state groups successfully conducts a large-scale cyber attack, competing non-state groups feel the need to conduct retaliatory attacks in response. These retaliatory attacks are not limited to just attacking each other but also include attacks against their respective governments as well. It is easy to see how this type of tit-for-tat action spurred on by nationalistic passions and lacking the formal constraints imposed by governments could quickly escalate out of control. If one of these groups successfully transitions to a kinetic cyber attack, other groups will attempt to follow and will likely try to do so in a more grandiose fashion in the spirit of one-upmanship. These types of retaliatory attacks could quickly lead to a shooting war in a region already dominated by nationalistic tensions and a history of violence.

There is some evidence that nation-states are utilizing non-state actors as proxies to conduct cyber attacks in order to achieve desired effects in or through cyberspace without the potential political recriminations that directly conducting such attacks would bear. Nation-states may overtly or covertly direct or support these non-state groups or, in many cases, simply turn a blind eye to their actions. Russia, in particular, seems to at least tacitly approve of the actions of patriotic hackers and the attacks these hackers have conducted against rival nations such as Estonia and Georgia, with no political recriminations for Russia. At the very least, Russia has failed to prosecute individuals that have been identified as having participated in these attacks and, in doing so, has failed to meet its responsibilities under customary international law. However, given the current debate as to if and how current international law applies to cyber conflict, much less non-state actors participating in cyber conflict, it is difficult to hold Russia and other nation-states accountable for these actions.

Legal considerations of cyber conflict

Currently, cyber conflict activities are mostly governed by the domestic laws, policies, and regulations of the respective states which undertake these activities. In the United States, for instance, cyber conflict activities by the military are largely governed under Title 10 and Title 50 of the US Code, which differentiates between military and intelligence activities and denotes who can perform such activities and under what circumstances and authorities they can be performed. While these titles do not specifically enumerate cyber conflict activities, they are applied to them as they are applied to military and intelligence activities in the other war-fighting domains. Domestic civil and criminal laws at both the federal and state level are also applied to activities in cyberspace, as demonstrated by the recent indictments under US criminal law of Chinese officials for hacking and cyber espionage.

At the international level, there is no specific body of law or treaty regime which governs cyber conflict. There is a great deal of debate as to whether international humanitarian law, sometimes called the Law of Armed Conflict (LoAC), applies to cyber conflict and, if so, under what circumstances. Opinions range from belief in the full application of the Law of Armed Conflict, implying that any use of force, regardless of the weapons employed, is covered by LoAC, to the opposite end of the spectrum whereby acts not explicitly forbidden in international law are generally permitted (NATO Cooperative Cyber Defence Centre of Excellence 2013).

In 2009, the NATO Cooperative Cyber Defense Centre of Excellence in Tallinn, Estonia, convened a group of international legal scholars to attempt to answer the question as to whether existing international laws applied to cyber issues at all and if so, how. The group agreed unanimously that in both *jus ad bellum* and *jus in bello* (legal concepts which describe the rights of a state to wage war and how a state should conduct itself during war), current international laws did apply, and the main effort of the Tallinn Manual details exactly how, in this group's estimation, existing law governs cyber conflict (NATO Cooperative Cyber Defence Centre of Excellence 2013). It is notable that the manual does not address those activities falling below the level of use of force, as the term is understood in *jus ad bellum*. 'Cyber espionage, theft of intellectual property, and a wide variety of criminal activities in cyberspace pose real and serious threats to all States. However, the Manual does not address such matters because application of the international law on *uses of force* and armed conflict plays little or no role in doing so' (NATO Cooperative Cyber Defence Centre of Excellence 2013). While the Tallinn Manual is thus far the best guide we have as to how international humanitarian law applies to cyber conflict, it is worth noting that this document has no legal authority and is simply opinion on the part of its authors.

An uncertain future

While there is already a great deal of activity in cyberspace that can be defined as cyber conflict, this is likely just the tip of the iceberg as activities associated with cyber conflict continue to evolve in the years to come. Nation-states around the world continue to develop and refine cyber warfare programs, and we are seeing a rapid militarization of cyberspace. This trend is leading to the development of very sophisticated cyber-exploitation toolsets and cyber weapons. State actors have focused heavily on developing the capability to stealthily infiltrate and maintain persistent access to their competitors' systems and networks. While this persistent access has thus far been used mostly for exploitation and the exfiltration of political, military, and economic data, it would be very easy to direct this level of persistent access toward even more disruptive purposes. Once an adversary gains access to a network, the difference between exploitation and disruption is no longer one of capability, but rather intent.

Conflict in cyberspace is not just centered on technical exploitation and disruption; there are also ongoing efforts by many states to shape the political, architectural, and regulatory environment at both the national and international levels. Governments in Iran, China, and Russia, among others, are attempting to control popular access to the Internet, regulate the information and content that crosses their state borders, and shape social networks to extend their control and influence. Similar efforts are ongoing in the international community to change the manner in which the Internet is currently governed to reduce US influence and potentially allow individual states to exercise more control over the content and architecture of cyberspace within their borders (Downes 2012). If successful, these changes could have severe implications for the open exchange of ideas and the network neutrality currently enjoyed in most of cyberspace.

The underlying architecture of cyberspace is also changing and evolving at a tremendous pace. New technologies are constantly being introduced, and existing technologies are converging to introduce network technologies into nearly every facet of life. Thermostats, televisions, refrigerators, automobiles, and wearable technology are being connected to the Internet. A new concept termed the 'Internet of Things' is emerging in which embedded technology in objects, animals, or even human beings are provided with unique identifiers and the ability to transfer data over a network without requiring human-to-human or human-to-computer interaction (Wigmore and Rouse 2014). Unfortunately, the evolution of these and other technologies is outpacing efforts to secure them, and this has serious implications for the future of both cyber conflict and information security at large. This rapid expansion and convergence of technologies is dramatically increasing available attack surfaces and introducing new vulnerabilities into devices and systems that were not previously accessible from the Internet.

While technology is rapidly evolving, legal and regulatory efforts are struggling to keep pace. At the state level, many nation-states have initiated legal and policy oversight to address these new technologies. However, in most cases, very little work has been done to directly address cyber conflict, instead concerned parties have relied on regulation of technologies and criminal prosecution of criminal activity. At the international level, there are currently no treaty regimes or international agreements which directly address cyber conflict. Some treaty organizations, such as NATO, are addressing the threats posed by cyber warfare through mutual-assistance frameworks. At the NATO Wales Summit held 4–5 September 2014, the organization endorsed new policies that put cyber defense in the same category as kinetic attacks. This change would imply that signatory countries would be required to come to the aid of member countries attacked by another country (Lemos 2014; North Atlantic Treaty Association 2014). Additionally, the United States and Australia recently added similar language to the Australia, New Zealand, United States Security Treaty (ANZUS) (Steward 2014). Yet, these agreements have yet to be tested in any meaningful way.

Cyber conflict will continue to be a widespread problem in the coming years. As these technologies evolve and are more fully integrated not only into everyday life but also into the military and intelligence communities of various nation-states around the world, policymakers will eventually be forced to address this phenomenon at the international level. Nation-states are already spending billions of dollars building cyber warfare and espionage programs in what may be the first great arms race of the 21st century. Additionally, the capabilities of non-state actors continue to grow and add tremendous complexity to an already challenging domain. Given the pervasive nature of cyberspace, it is vitally important that the international community begin to address some of these issues and build consensus on what is or is not acceptable behavior in this domain. Failure to do so may have dire consequences.

3
Establishing Cyber Warfare Doctrine[1]

Andrew Colarik and Lech Janczewski

Introduction

Over the past several decades, advances in technology have transformed communications and the ability to acquire, disseminate, and utilize information in a range of environments. As a result, modern armies have advanced their command-and-control capabilities by using a robust information space through network-centric warfare. The ever-increasing convergence of military and commercial operations warrants considering the possibility that communication and information infrastructures are viable components – both as targets and weapons – in times of war.

Developments in recent years indicate that Internet and communication technology (ICT) in particular are becoming a viable theater of military conflict. The possibility of widespread conflicts fought in cyberspace continues to arise as digital-warfare capabilities are developed. The deployment window for a cyber attack has a dramatically different form from traditional conflicts and thus requires a different planning defense structure. Such an attack could be quickly prepared by a relatively small group, launched without warning from anywhere on the globe against any possible ICT target, and escalate in a matter of minutes to shut down national infrastructures (Parrish 2011). In this context, each modern state should be prepared to be the target of a cyber warfare attack and stand ready to launch a counteroffensive. Preparations for such conflicts have already started in many other countries, including Israel, North Korea, Iran, and Russia (The Economist 2010). When we examine these activities from a more holistic perspective, the preparation for both offensive and defensive cyber capabilities has both technical and public-policy components. In other words, nations need to find answers and solutions to questions such as the following:

- What activities must be undertaken in the case of a cyber attack against a nuclear power plant?
- What is the measured and appropriate response to such an attack?
- What level of attack threshold constitutes an act of war?

With cyber attacks, there is no time to deliberate a comprehensive response. We believe modern nations lack a grand strategy for handling cyber attacks, one that gathers and coordinates their national resources for shared security and prosperity (Liddell-Hart 1967). Hence, we suggest that each country develop a cyber warfare doctrine (CWD) that includes all stakeholders, brings about a decisive conclusion when such attacks occur, and serves to deter future conflicts through a unified national security policy.

Developing a comprehensive CWD is a complex task requiring much preparation. Nevertheless, both civilian and military establishments have made considerable progress toward securing their national infrastructures and preparing for war in the cyber realm. Unfortunately, these efforts are being developed and implemented in a piecemeal manner. The planning components of both civilian and military interests are separate and disjointed, regardless of provisions that both sides believe will create synergistic outcomes (The White House 2011). What's missing is a true systems approach to handling conflicts originating in cyberspace that cross many jurisdictional boundaries and interests. What's needed is a general national policy on how to handle IT-based attacks that disturb a country's normal functioning. Such a policy should embody a set of self-defense principles inclusive of civilian infrastructure, military objectives, and national security policy. This article argues for establishing a CWD that would be used to determine a nation's appropriate response when attacked via cyberspace. Such a doctrine would be used as a guide for defense forces in a time of conflict; as a unified governing philosophy for military operations, deployment to protect civilian infrastructure, and the governance of international cyber relations; and as a deterrent to future adversaries.

The objective of this article is to summarize the considerations that would allow senior leadership to develop a comprehensive, strategic CWD. We will discuss the justification for considering information technologies in military conflicts and the events supporting our supposition, the various doctrines that will form the basis for developing a CWD, and the possible components of a CWD. We will also propose a national collaborative framework for obtaining stakeholder buy-in for a CWD and offer some final conclusions.

Setting the stage: milestones in cyber war

In his celebrated book *On War*, Carl von Clausewitz defines war as merely a duel on an extensive scale (Clausewitz 1976). While such conflicts commonly occur between two parties, sometimes they enlarge to encompass multiple states, regional alliances, and federations of nations, in which the conflict is essentially between two sides; for example, the Axis Powers and the Allies in World War II. These types of wars, generally called symmetric, are often characterized by large conflicts between sides of relatively equal strength, resources, and technological capabilities. For many reasons – such as international treaties, global governance initiatives, and advances in military technology – these duels are becoming less frequent. Two of the biggest conflicts in recent years, in Iraq and Afghanistan, started as duels but quickly migrated toward a different type of conflict: asymmetric war.

In its simplest form, 'have-nots' undertake such warfare against 'haves,' seeking victory by employing their specific advantages against the vulnerabilities of a much stronger opponent (Thornton 2007). To overcome the disparity in strength, the weaker antagonist looks for asymmetric options, targeting the 'will' of the stronger opponent by strategically pursuing disproportionate psychological effects (McKenzie 2001). In the authors' view, being able to turn opponents' greatest strengths into their greatest weaknesses is the highest, most refined form of asymmetric warfare.

For decades, adversaries have compromised the use of information technology by stealing financial, proprietary, and/or secret information, and they continue to do so today (Stoll 1989; Alperovitch 2011). So dependent on technology have modern nations become that they are fundamentally weakened when such systems and processes are disrupted for any meaningful time. This vulnerability, of course, continues to have national security implications. As a result, numerous national and international efforts have been made to develop policies for combating the use of cyberspace for criminal activities (Schoeff 1998; Council of Europe 2001; The White House 2003; European Parliament 2004; Eijndhoven 2011).

In recent years, a series of milestones have provided clear indicators for the viability of asymmetric conflicts originating in cyberspace. The most significant of these have involved the European nation-state of Estonia, which has implemented a highly integrated e-government infrastructure. The country was forced to digitally isolate itself when unknown, politically motivated attackers initiated a series of distributed denial of service (DDoS) attacks incorporating one-million-strong botnet. The attacks on Estonia's Internet systems began in April 2007 and lasted three weeks, but it is their sustained impact that's of primary importance to this discussion rather than the technological methods employed (Ruus 2008). We believe the attacks, which the Estonian government has labeled 'cyber terrorism,' have provided the impetus for nation-states to deploy cyber-offensive capabilities for future conflicts.

Another milestone were the cyber attacks on the former Soviet-bloc state of Georgia, which mimicked the events that occurred in Estonia. The country's national communication infrastructure was shut down just before Russian military forces entered its borders in August 2008. Government websites and news outlets as well as banks, including Georgia's largest, were affected. While the country's use of the Internet is still emerging, the effects of the attacks limited the government's ability to spread its messages in time of crisis. Efforts to engage and perpetuate the cyber attacks during this period were consistently conducted in Russian, but no verifiably responsible party has been identified to date (Tikk, Kaska and Vihul 2008).

The last milestone relates to the first detection of a new breed of computer worm known as Stuxnet, which appeared in June 2010. It is considered the first worm specifically created to target real-world infrastructure, such as power stations and water plants. After it has hijacked a PC, Stuxnet looks for Siemens software that runs industrial control systems and begins to speed up or slow down power generation for destructive means, which was the case for Iran's Bushehr nuclear plant. The worm's resemblance to legitimate software, such as digital certificates, while using a self-launching, zero-day vulnerability in the attack, allowed its rapid, unobstructed distribution through the *a priori* assumption of security software that if a program meets certain conditions, it is trustworthy (Mastrosov et al. 2011).

The ramifications of the above milestones to national infrastructures in times of conflict are staggering. Not only can a nation's communication channels be disrupted as a force multiplier, but basic infrastructure, such as power and water distribution, can also be remotely attacked and disabled, putting the targeted country at a distinct disadvantage. The military establishments of many countries have for some time recognized the possibility of a cyber war, and we believe the above milestones were the impetus for both the viability and necessity of recent cyber mobilizations. A 1996 report prepared for the US Office of the Secretary of Defense projected that battlefield command, control, communications, and intelligence (C3I) vulnerabilities 'may become less significant than vulnerabilities in the national infrastructure' (Molander, Riddle and Wilson 1996).

In June 2009, the US Cyber Command was created, and in July of 2011, Deputy Secretary of Defense William J. Lynn III announced that as a matter of doctrine, cyberspace will be treated as an operational domain similar to land, air, sea, and space (US Department of Defense 2010; Lynn 2011). In their notable book, *Unrestricted Warfare*, Chinese colonels Liang and Xiangsui claimed advanced technology gave the country's adversaries a significant advantage and proposed that China 'build the weapons to fit the fight (Liang and Xiangsui 1999). Recently, the Chinese People's Liberation Army (PLA) confirmed the existence of its Online Blue Army (Lebowitz 2011). Many smaller countries have also begun developing cyber warfare divisions but thus far have kept such capabilities plausibly deniable or at a low profile to avoid preemptive reprisals (Fogarty 2011).

All of the above are major indicators that information technologies are already playing an important role in carrying out military objectives. These examples are but a small collection of high-profile cases, and we believe sufficient evidence points toward countless lower-profile events that have gone unreported or have been classified. In the next sections, we will discuss the theoretical foundations of national doctrine in preparation for addressing the larger strategic importance of technology's role in achieving military objectives.

Doctrine

In its simplest form, a doctrine is defined as a body of principles that form a system of belief. A doctrine can be considered a statement of fundamental government policy, a principle of law established through past decisions, or a military principle or set of strategies (Webster Dictionary 2002). In essence, a doctrine embodies the rules by which individual societies govern themselves and maintain standards. Therefore, a CWD, in principle, represents a set of rules and standards for governing a war involving cyberspace. When viewed in more depth, a doctrine brings with it a set of characteristics stemming from the people who embrace it and their societal processes. Doctrine functions to provide a tempered analysis of experience and a determination of beliefs, teach those beliefs to succeeding generations, and offer a common basis of knowledge and understanding that can provide guidance for action (Drew and Snow 1988). In other words, doctrine states how to do something the best possible way and is passed on to subsequent generations. This implies that it is based on knowledge accumulated from making strategic decisions within a domain. Therefore, doctrine may take many forms, either fact dependent and limited in scope or broadly interpreted and sweeping in its breadth of application (Tiller and Cross 2006).

To form a CWD, the dominant doctrines governing a people's lives must be examined. The authors believe a CWD must reflect the accepted doctrines governing how we organize and direct ourselves, interact with others, pursue prosperity, and defend ourselves collectively if future generations are to sustain and support it. In the following section, we shall briefly examine three dominant doctrines. The first is political, which is most often reflected in a nation's constitution or its equivalent. The second is legal, which often follows political doctrine as an embodiment of societal interaction. The last is military, which governs how a nation secures itself.

Political doctrine

Throughout much of the West, it is traditionally understood that open, democratic societies must have a set of common understandings or governing principles that allow people to mutually interact. These principles are articulated through political doctrine, which establishes the political and social structure of a nation and the supporting structures that grow from its principles and shape the doctrine's impact on future generations. An example of political doctrine is a constitution, which may be considered a set of fundamental principles, or established precedents, according to which a state or other organization is governed (New Oxford Dictionary 2005).

We believe the political doctrine embodied in a nation's constitution articulates the will of a people, a new order, a new way of governance, or a new social doctrine. It is in this document that the prevailing political principles for governing reside and are used to shape a society's social and legal environments. A national constitution is usually a mirror of a nation's past and a prescribed program for its future. An example of such a document is the US Constitution, adopted in 1787 at a critical moment in the country's history. Major conflicts with Spain and England remained unresolved, large amounts of capital had been drained by the war of independence, and a substantial number of states were unsure about union membership. The country needed an efficient system of national government while maintaining individual states' rights. The Constitution's preamble reflects this by stating: 'We the people of the United States, in order to form a more perfect union, establish justice, insure domestic tranquility, provide for the common defense, promote the general welfare, and secure the blessings of liberty to ourselves and our posterity, do ordain and establish this Constitution for the United States of America.' The document was created at a pivotal point in history, when American society was compelled to choose the manner in which it would govern itself. We believe that to be consistent with a society's values and internal processes, a CWD must embrace its principles, as illuminated in a document such as a constitution.

Legal doctrine

Once it embraces a political doctrine, a nation must invoke systems and processes that exemplify its 'spirit.' These are often articulated in its legal doctrine, considered the currency of the law, in that established precedent becomes the foundation for determining the application of law in future cases. In law, rules tend to be strict requirements that identify the answer to a dispute once the facts have been established, while standards are more like guides for resolving disputes after identifying a set of factors to be considered and balanced.

One of the dominant doctrines in law is *stare decisis*, Latin for 'let the decision stand.' This is a legal principle by which judges are obliged to respect the precedents established by prior decisions. This overarching principle guides courts in following standards established by decisions in earlier cases (Kozel 2010). Thus, judicial activism is minimized and consistency in future judicial rulings is established, allowing a people to better understand their current and future legal obligations when interacting with one another. With *stare decisis* in mind, we strongly suggest that any new doctrine, including a CWD, must be based on a country's current legal doctrine and established precedents. There may be times when new precedents must be established – for example, when technological developments impact politics and the economy – but while they may help a nation better handle recurring problems, they may also create unforeseen consequences and dysfunctions between the law and its enforceability. The authors therefore caution the framers of a CWD against forcing new precedents without fully considering their possible consequences.

Military doctrine

Once a nation has decided on its basic political and legal doctrines, it must address how it intends to defend itself and its chosen way of life. Military doctrine embodies the fundamental principles by which a country's military force guides its actions in support of national objectives (Department of Defense 2010). This doctrine can in turn be divided into fundamental, environmental, and organizational doctrines, which identify key military factors and address how each will be governed and under what conditions. The nature of war, as well as the purpose of military forces and their relationship to other instruments of power, reside in fundamental doctrine, which is relatively insensitive to political philosophy or technological changes. The following examples are typical statements of fundamental doctrine:

- War is the failure of policy.
- The object of war is to overcome an enemy's hostile will.
- The object of war is a better state of peace.

The fundamental portion of a CWD may include such statements as governing principles in developing environmental and organizational components. The cumulative understanding of military deployment in a particular operating medium – such as sea, air, land, and space – forms environmental doctrine. In the case of a CWD, the environment is likely to embody information technologies and computer networks and their physical infrastructures. Organizational aspects likely will be adaptations of existing structures. Both environmental and organizational components will require consideration throughout the CWD development process.

Developing CWD components

Considering the many details critical to formulating a CWD, serious consideration of a collaborative and discussion framework is crucial. This section offers a conceptual starting point and presents several key strategic questions we believe essential to forming a cyber warfare doctrine. We offer the following questions as critical examples of the types of queries needed to enable a corrective and decisive response following an attack.

What is the line between cyber warfare and traditional warfare?

Definitions matter when implementing policy, and in developing a CWD a variety of factors must be considered. In essence, this question focuses on the role of information technology as an enabler of warfare and, therefore, as a viable target from both attack and defense viewpoints. We believe cyber warfare will have kinetic world effects, meaning it will cause real direct and indirect damage to physical infrastructure (Parks and Duggan 2001). The notion that an information-age war would be bloodless and sterile is challenged by the fact that our digital infrastructures and physical capabilities are integrated in order to sustain and support modern warfare (Sullivan and Dubik 1994).

Information is the central element for commanding the conflict space; of equal importance is the infrastructure that allows information to flow. From a strategic perspective, we must consider an opponent in a cyber war as a system composed of a series of subsystems (Warden 1995). Each subsystem that supports and sustains the larger system enables a country to direct its resources toward the conflict. The essential question before us is this: To what extent will cyber warfare conflict encompass real-world assets and lead to full-scale war? Without a clear distinction between cyber and traditional warfare, how can any country in the modern world take action that is justified, rational, and proportional to a given attack? We believe such a distinction is crucial.

What is the CWD conflict space?

Controlling the conflict space is central to resolving military conflicts. But when a country seeks to do so, what exactly are its objectives? The authors believe cyber warfare involves preventing opponents from knowing anything about you while you seek to know everything about them (Arquilla and Ronfeldt 1999). This 'knowledge battle' extends to a nation controlling its own resources while rendering its opponents ineffective.

Preventing the use of resources is central to controlling the conflict space. In an increasingly interdependent global economy, the implications can quickly escalate to encompass unforeseen consequences. Battle space dominance likely will include a nation's information infrastructure as well as the information flowing through it from both sides of the conflict. We must also consider all of the pathways and infrastructure between the two parties in a conflict, as information infrastructures are rarely symmetric. These third-party pathways likely would be active or unwitting participants in attack and defense measures, as their infrastructures would support such activities. Because of the distributive architecture of cyberspace, defining the conflict space both in totality and in conflicts as they arise is the first step in developing strategies to control it.

What threshold of aggression constitutes the level of CWD response?

Waging a moral war is essential to sustaining it, and we believe a cyber war is no different. A measured response to a cyber warfare attack requires deterrence and escalation levels (Tirenin and Faatz 1999). The strategic objective in a conflict is to cause an opponent's systems to change to such an extent that it is forced to adopt your objectives or become incapable of mounting an opposition. In a retaliatory action aimed at deterring future attacks, a CWD should establish responses that would cost the aggressor more than it might stand to benefit from such attacks, thereby encouraging restraint. If a significant penalty beyond an incident is not established, escalating attacks likely will occur. But to remain moral, such responses must be proportional. Thus, an assessment of any attack should be accompanied by a suitable and corresponding response to maintain its moral justification when damages or casualties are incurred. Therefore, determining the range of responses, along with their alignment with a nation's strategic security goals, is critical to responding responsibly to attacks.

What is the definition of a CWD victory?

War must be waged with a constant regard for the peace desired. As with a response and escalation policy, knowing what constitutes a CWD victory is essential to not overreaching. Responses to aggression must consider taking possession of the opponent's strengths as well as destroying its armed power, all while considering public opinion. Levels of desired outcomes must be tied to any attack response. Depending on the assault's severity, victory may be limited to restoring operations and taking steps to improve defensive measures. The prevention of future attacks would then be a consideration in establishing victory conditions. In larger-scale conflicts, the partial or complete disabling of an aggressor's attack capabilities may be warranted and considered a victory condition. When an aggressor has the full support of the country, the elimination of the attack infrastructure may be warranted, thus also serving as a victory condition.

The authors believe that a clear understanding of victory is crucial to the formation of a CWD, as it has profound policy implications for future peaceful relations with both the antagonist and the rest of the world.

What are the principal CWD assets needed to win?

A comprehensive understanding of the assets needed to wage a cyber war is essential to creating the infrastructure for supporting a CWD. Identifying the dependencies a military force has on information technology is tightly coupled with defining the needed assets. Modern militaries rely heavily on systems that provide speed of command in order to achieve information superiority and the massing of effects. The result of the rapid foreclosure of enemy action and the shock of closely coupled events makes network-centric operations a significant strategic advantage (Cebrowski 1998). Assets that enable increased information richness, reach, and shared awareness are responsible for the transformation of improved awareness into collaborative planning and synchronized action (Alberts et al. 2001). Assets that provide for peacekeeping measures, such as border management and verification activities, play a role in threat awareness and removal (Cahill, Rozinov and Mule 2003). We believe these assets, and any suitable measures to defend or attack them, must be identified to prevent a CWD from lacking the proper scope and depth when implemented.

What are the factors inhibiting an effective and decisive CWD response?

We live in a global community of communities, and the more integrated and interdependent the world becomes, the more our policies and responses will be moderated by those indirectly connected to our actions. Cyberspace is made up of core national and international infrastructures residing in a multitude of legal jurisdictions and global alliances. We believe that without fully considering the regional and global consequences of taking direct action in a cyber war, no CWD would bear a substantive relation to a nation's larger strategic security policy. By understanding this sphere of influence and articulating the implications, profound long-term ramifications can be mitigated and a greater understanding of state actors can be cultivated. The authors believe that a CWD should contain remedies for factors that would inhibit effective and decisive responses in any cyber war conflict. While additional issues undoubtedly will be raised while forming a CWD, we think the queries presented here are representative enough, in both depth and scope, to illustrate key elements that must be considered in such an endeavor.

Collaborative and discussion framework

The foundation of a nation's information infrastructure is generally distributive in nature, and the creation of a CWD is no simple task if it is to include its closest stakeholders. Drawing upon the past for guidance, the authors propose a collaborative and discussion framework similar to the one used by President Dwight Eisenhower in confronting the expansion of communism and the threat of nuclear war. The president believed the best way to formulate national policy in a democracy was to assemble the best-qualified people with opposing views on an issue and listen carefully as they debated it (Eisenhower Memorial Commission 2011). This approach formed the foundation of Project Solarium, which resulted in a doctrine that governed the Cold War, and whose effects are still felt today (Eisenhower 1953).

We believe several fundamentals must be observed for a collaborative and discussion framework to be successful. The first of these is that a CWD initiative should be originated and governed by top government officials, as executive government is in the best position to facilitate and coordinate such an endeavor. Second, participating experts should be delegates from civilian government, defense, security, and professional organizations related to information technology, so that a broad set of skilled stakeholders are represented in the problem-formulation and -solving processes. Third, a relatively short time should be allocated to creating a CWD, forcing participants to stay focused on the tasks at hand and not expand the mandate's range and scope. Fourth, the results of such an endeavor should be accepted by the nation's head of state and be widely published. This last step is critical, for the CWD's dissemination establishes new norms of conduct and consequences for their breech. This doctrine transparency is in keeping with the highest traditions of open, democratic societies and clearly changes the rules of digital importance as a national security imperative.

The proposed framework has three phases. In Phase One, the head of a nation initiates the CWD collaboration process by identifying and selecting the primary stakeholders – from business, government, and professional organizations – who have both a vested interest in any CWD outcomes and the expertise to contribute substantively to the endeavor. Each organization selects delegates to represent it at a convention charged with creating a key set of questions such as those contained in the previous section.

Once these questions have been agreed on, Phase Two commences, with the questions being sent to the selected organizations in Phase One for deliberation. Because knowledge is often held by unlikely participants, and due to the larger implications of a CWD, the authors strongly suggest that these hearings be held in public forums, at which constituents may freely offer policy ideas in answering the key questions.

Once assembled by participating organizations, these policy suggestions form the basis of Phase Three, in which the business community, government branches and agencies, and professional organizations jointly assemble before the nation's head of state and Security Council to present their ideas for answering the questions and to defend their proposals against opposing viewpoints. Critical to this phase is that all

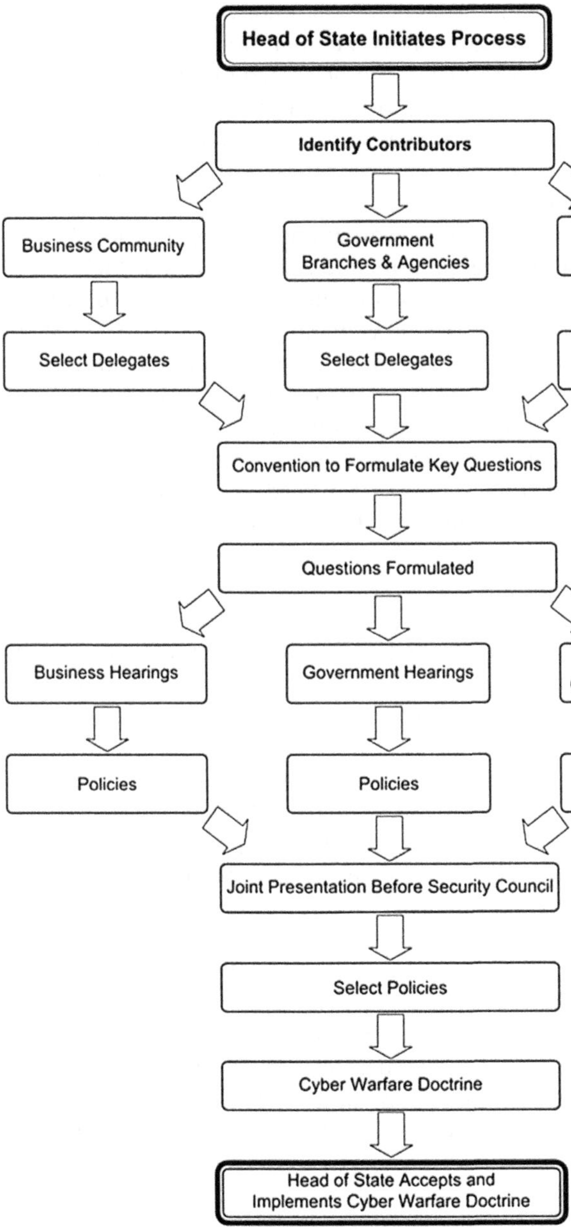

Figure 3.1 Collaborative and Discussion Framework

participants rigorously review the areas of discourse. It falls to the Security Council to assemble policies common among stakeholders as well as those that have withstood rigorous examination by all participants. A final document is then formulated and presented for ratification.

The authors believe the above process would create a CWD that has stakeholder input and buy-in, that addresses a nation's main cyber war concerns, and that permits decisive action with the backing of the nation-state (see Figure 3.1).

In summary, the proposed framework for developing a national CWD is based on several fundamental principles. The first is that such a venture would be conducted at the head of state's discretion, regarding both initiating the CWD process as well as its final acceptance. The CWD's development would be delegated to specialists from civilian government, defense personnel, security, and professional organizations related to information technology. The final proposals would be presented to the national-level Security Council (or other body with similar responsibilities) and ultimately be accepted by the head of state. The final CWD formulation would then be placed in the public domain.

Conclusion

Information technology has reached a level of development and integration into modern societies that allows it to be used to damage a nation's well-being. Numerous examples are available, and this article has presented several of them. Attacks on information systems and infrastructures may soon escalate into a full-scale military conflict. Whether such a confrontation is provoked by third-party cyber criminals or state-sponsored forces, a country would do well to be prepared. Many other defense forces are also developing or mobilizing themselves for cyber conflicts on a national and international level. To our knowledge, no state to date has a comprehensive national strategy for handling a cyber war that aligns the civilian infrastructure with military operations in a collaborative environment. In this article, we have summarized many of the dominant issues that must be addressed to formulate a comprehensive national CWD. We have outlined a collaborative process that brings together the government, business, and professional organizations responsible for a country's cyber infrastructure and national security.

Note

1 This chapter has been published previously in the *Journal of Strategic Security* Volume 5 Issue 1 2012: 31–48.

4
How Cyber Changes the Laws of War[1]

Jack Goldsmith

Introduction

Michael Walzer's *Just and Unjust Wars* anticipated many problems and developments in the laws of war. But it understandably did not anticipate how the Internet and associated computer and telecommunications revolutions would change war or the laws that govern it. In 1977, the year Walzer's book was published, Arpanet, the Internet's precursor, had been operating in practical secrecy with a crude packet-switching system for just eight years; the first e-mail system was five years old; and Vinton Cerf and Bob Kahn had used the term 'Internet' for the first time in a paper on the transmission control protocol just three years earlier. No one at this time, or for more than a decade, would worry much about the internal security vulnerabilities of this developing communications system. The wake-up call for security, such as it was, came in 1988, when Robert Tappan Morris, a graduate student at Cornell, introduced a worm on the Internet that was designed to determine the Internet's size but that inadvertently shut down about 10 percent of the 60,000 computers then connected to it (Orman 2003). This event startled the Defense Advanced Research Projects Agency (DARPA), the futuristic Department of Defense (DoD) research wing. DARPA had given its financial support to what became the Internet to ensure that military communications could withstand nuclear attack. But suddenly its young creation seemed vulnerable from within.

These vulnerabilities would grow in the next two decades because, despite growing cyber-security concerns, the military and the society it defends would become ever more reliant on ever more vulnerable computer and telecommunication systems. This article explains these vulnerabilities, sketches how they affect the laws of war, and conjectures that international norms to regulate and temper attacks on these vulnerabilities are unlikely to develop.

Characteristics of cyber

Many factors make computer systems vulnerable, but the most fundamental factor is their extraordinary complexity (Goldsmith 2010). Most computers connected to the Internet are general-purpose machines designed to perform multiple tasks. The operating-system software that manages these tasks, as well as the computer's relationship to the user, typically has tens of millions, and sometimes more than 100 million, lines of operating instructions, or code. It is practically impossible to identify and analyze all the different ways these lines of code can interact or might fail to operate as expected. And when the operating-system software interfaces with computer processors, various software applications, web browsers, and the endless and endlessly complex pieces of hardware and software that constitute the computer and telecommunications networks that make up the Internet, the potential for unforeseen mistakes or failures becomes unfathomably large.

The complexity of computer systems often leads to accidental mistakes or failures. We have all suffered computer crashes, and sometimes these crashes cause serious problems. In 2009, the Internet in Germany and Sweden went down for several hours due to errors in the domain name system, which identifies computers on the Internet. A few years ago, a software problem in the Pentagon's global positioning system network prevented the Air Force from locking onto satellite signals on which they depend for many tasks. The accident on the Washington Metro in 2010, which killed nine people and injured dozens, was probably caused by a malfunction in the computer system that controls train movements. Several years ago, six stealth F-22 Raptor jets on their maiden flights were barely able to return to base when their onboard computers crashed.

The same complexity that leads to such malfunctions also creates vulnerabilities that human agents can use to make computer systems operate in unintended ways. Such cyber threats come in two forms. A cyber attack is an act that alters, degrades, or destroys adversary computer systems or the information in or transiting through those systems. Cyber attacks are disruptive activities. Examples include the manipulation of a computer system to take over an electricity grid or to block military communications or to scramble or erase banking data. Cyber exploitations, by contrast, involve no disruption but refer to merely monitoring and related espionage on computer systems as well as the copying of data that is on those systems. Examples include the theft of credit-card information, trade secrets, health records, or weapons software and the interception of vital business, military, and intelligence communications.

Both cyber attacks and cyber exploitations are very hard to defend against. 'The aggressor has to find only one crucial weakness; the defender has to find all of them, and in advance,' wrote Herman Kahn in his famous 1960 book, *On Thermonuclear War* (Kahn 2007). This generally true proposition about defense systems has special salience for computer networks. Even if (as is often not the case) those trying to find and patch computer vulnerabilities outnumber those trying to find and exploit the vulnerabilities, the attacker often still has an advantage. Under the Kahn principle, in some fraction of the time the attacker will discover a vulnerability that the defender missed. And she need only find one, or a few, vulnerabilities to get in the system and cause trouble.

Once a vulnerability is identified, an attack or exploitation is relatively easy to disguise, because the operation of a computer is almost entirely hidden from the user. Malware can be embedded in a computer system without the user's knowledge, either remotely (when the user downloads an infected program or when she visits an infected website) or at any point in the multi-country global supply chain that develops and produces most commercial software. And once it is embedded, malware can be used for any number of tasks, including data destruction, theft, taking over the computer for various purposes, recording keystrokes to discover passwords, and much more. Many forms of malware are hard for engineers to find through diagnostic testing and are missed by antivirus software. Computer users often do not discover malware before an attack makes clear that something has gone wrong. They often never discover malware that facilitates computer exploitations or (as happened in the China-Google kerfuffle), they discover it too late.

The inherent insecurity of computer systems is exacerbated by the number and incentives of actors around the globe who are empowered to take advantage of computer vulnerabilities. In real space, geography serves as a natural barrier to attack, theft, and espionage: only if you get near the Pentagon can you attack it; only if you get near the Citibank branch in New York can you rob it. And if you are near these places in real space, American law enforcement and military authorities can exercise their full powers, within US sovereignty, to check or deter the attack. In cyberspace, geography matters much less because the Internet links computers globally with nearly instantaneous communication. As

China's recent theft of information on Google's proprietary computers shows, someone sitting at a terminal in China can cause significant harm in the United States. And of course there are countless people around the globe with access to a computer who would like to do bad things inside the United States. To the extent that they are located outside the United States, American law enforcement authorities have much less effective power to stop or deter them. The FBI must rely on law enforcement authorities in foreign countries, who are often slow and uncooperative, giving bad cyber actors time to cover their tracks. And the American military cannot enter a foreign country unless the threat or attack rises to the level of war (a topic to which we will return).

Law enforcement and military authorities seeking to check malicious cyber activity face another fundamental challenge: the 'attribution problem' of identifying the author of a cyber attack or cyber exploitation. It is very difficult, and very resource-intensive, and sometimes impossible, to trace with much certainty the computer origin of a professional cyber attack or cyber exploitation; it is even harder to do so in real time or even in the short term. A thoughtful adversary can hide its tracks by routing attacks or exploitations through anonymizing computers around the globe. In 2009, a denial of service attack – a massive spam-like attack that clogs channels of communication – brought down some American and South Korean websites. Early reports said that the attack came from North Korea, but a few weeks later it was learned that the attack originated in Miami (and possibly, before Miami, elsewhere) and was routed through North Korea. It is still not known for sure who launched the attack, or from where.

Even if we can determine with some certainty which computer in the world is behind an attack or exploitation, that fact alone does not indicate who, or even which country, is responsible for the aggression. In 2009, a detailed study by the Information Warfare Monitor uncovered an extensive plot known as 'GhostNet,' which emanated from computers in China and infiltrated more than 1,000 sensitive government and commercial computer systems from over 100 countries. But the report could not determine whether the plot was controlled by the Chinese government, or by private 'patriotic hackers' acting in the Chinese interest but without government involvement, or by a criminal network in China. Nor could it rule out the possibility that 'a state other than China' was behind the plot, using agents to launch the operation from China in an attempt to 'deliberately mislead observers as to the true operator(s) and purpose of the *GhostNet* system' (Information Warfare Monitor 2009). It is still not known who is behind the Conficker worm or the July 2009 denial of service attack on South Korea and the United States. Nor, more recently, do we know for sure who is behind the Stuxnet worm. Law enforcement and military officials are hobbled not only by geography, then, but also by their inability to know for sure where and by whom a cyber attack or exploitation originated.

To date, most harmful cyber operations have taken the form of exploitation – espionage, and massive theft of intellectual property, military secrets, and the like. But cyber operations can also be attacks that potentially rise to the level of war, or that facilitate war-fighting. We saw glimpses of this when Russia (or groups in Russia) used denial of service attacks to shut down Estonian banks and government websites in 2007 and cripple Georgian government websites in 2008. Presidents Obama and Bush reportedly ordered covert computer attacks on computer systems related to Iran's nuclear weapons program. Many experts believe that a cyber operation could shut down a stock exchange or destroy bank or money-transfer records or operations, wreaking economic havoc. Or a cyber operation could corrupt or take over the computer systems (known as SCADA systems) which monitor and control infrastructure processes like the electrical grid and nuclear power plants, causing them to shut down or malfunction. The significance of the Stuxnet worm is that it seems designed to do just that.

Cyber and the laws of war

Taken together, the factors outlined in Section 1 make it much easier than ever for people outside one country to commit very bad acts, possibly rising to the level of war, against computer systems and all that they support inside another country. This raises some well-studied (though not resolved) challenges to the laws of war, and some less obvious ones.

One challenge is to figure out when a cyber attack implicates *jus ad bellum*. The hard question is how to translate the UN Charter concepts of 'use of *force*' and '*armed* attack' into the cyber realm. The main answer that has emerged, drawing on Michael Schmitt's work, has been to focus on the scale of the kinetic effects of the cyber operation (Schmitt 1999, 2010). When the effects of a cyber operation are akin to the effects that would implicate the UN Charter terms, then cyber operations implicate the UN Charter. So, for example, a cyber attack that renders the electricity grid or air-traffic control system inoperable, and that as a result causes many deaths, would count as a use of force. But a cyber operation that merely involves espionage or that disrupts DoD computers conducting military research likely would not be considered a use of force.

These cases are easy enough. But cyber operations introduce more challenging questions (Barkham 2001; NRC 2009; Waxman 2011). The challenges arise mainly because the Charter focuses its prohibitions on military means of inflicting damage on another state but does not prohibit economic or political means of inflicting damage on another state. As a general matter, military means by one state that leads to deaths or physical destruction in another implicate the Charter, but political or economic sanctions that lead to deaths or physical destruction in another state do not. Cyber operations can cause havoc in a nation, including death and destruction, which might appear more like economic sanctions than a military use of force. Consider, for example, a cyber attack that corrupts data on a stock exchange and which in turn causes widespread economic harm but no direct physical damage. Is this more like a physical use of force or like economic sanctions? What about widespread economic harm caused by massive theft of digitalized intellectual property? Theft and espionage are not generally viewed as implicating the Charter, but the cyber context changes the scale and consequences of theft and espionage to a degree that can result in harm to the country at least as severe as a physical attack. Another difficulty with cyber operations is that, unlike many kinetic attacks, they can take place slowly and can be reversible. There is no settled answer to the question of whether or when a slow disruptive cyber attack on critical infrastructure or an analogous system that gradually renders it sub-optimally operable becomes a use of force. Similarly, there is no settled answer to the question of whether a temporary but reversible shutdown of a computer system, lasting perhaps two days or two weeks, associated with a fighter-jet squadron or a reconnaissance-satellite system, is a use of force. Nor is it clear whether 'mere' destruction of critical economic or military data, without any physical consequences, is a use of force.

Similar questions arise in trying to figure out which cyber attacks are 'armed attacks' under Article 51. In addition to conceptual problems analogous to those that arise with uses of force, the problem of attribution causes further complications. A thoughtful adversary can hide its tracks by routing attacks through anonymizing computers around the globe. Even if a nation knows which computer in the world is behind an attack, that fact alone does not indicate who, or even which country, is responsible for the aggression. Sometimes traceback and related forensic tools can provide pretty good attribution. And human and other forms of intelligence gathering can further help in attribution. But even taking into account these and other tools, the attribution of a sophisticated cyber attack is neither fast nor remotely certain. This makes it very hard for the nation responding to an armed attack to know which nation (if any) is responsible for it, and thus which nation it should use force against in self-defense. Opportunistic but plausible denials of responsibility for the armed attack will be frequent, attribution assessments will be probabilistic, and mistakes in responding to cyber attacks will be inevitable. These problems create disincentives and uncertainty in responding to a cyber attack, and they lower *ex ante* disincentives to cyber attack in the first place. A related problem is that a cyber attack might start slowly and build over time, and waiting too long to respond to it might well make it harder to respond. But there is no way to know in advance whether an attack will grow in this way. Must a nation wait until the attack crosses a critical threshold when it might be too late, or can it respond earlier even though its prediction about the ultimate scale of attack might be faulty?

The attribution problem also underlies many of the problems in applying *jus in bello* principles. Cyber operations challenge both the principle of distinction and the principle of proportionality. In the cyber realm, it is often hard to know whether the computer system being attacked, or the nation associated with that computer system, is a military target. The mingling of civilian and military computer and telecommunication systems raises a similar problem. These examples suggest that a nation using cyber weapons often, and perhaps usually, cannot know with certainty whether it is attacking a military or civilian target. (Similar problems can arise, obviously, in the kinetic context, but the problem is much more pervasive when cyber weapons are employed.) In addition, a cyber attack can have unpredictable indirect and cascading effects on associated computer networks that make collateral damage very hard to calculate – much harder than the vast majority of kinetic targeting decisions. One reported reason why the Bush administration called off a planned attack on Saddam Hussein's financial network in 2003 was a worry about uncontrollable indirect effects on the global banking system (Markoff and Shanker 2009).

The laws of war are somewhat muddled when the source of an attack from one state is a non-state actor. The problem is exacerbated in the cyber realm. For one thing, a nation suffering an attack has a hard time knowing whether the attack comes from a state actor or a non-state actor. This can make it hard to know whether a military or law enforcement or some other response is appropriate. Assuming the actor is private, the criteria for state responsibility are unsettled. There are growing calls to deal with this cyber-attribution problem by making a nation responsible for all cyber attacks that emerge from within its borders, even if the attacks are not sponsored by that nation (Clarke and Knake 2010). This would in theory ameliorate the attribution problem by eliminating the 'it wasn't us, it was private hackers' defense that Russia, China, and other nations have invoked when criticized for cyber attacks from inside their borders. It is not clear whether technology permits nations (in a remotely cost-effective manner) to take the steps needed to control or arrest all malicious cyber agents from leaving their borders. But assuming away the technological hurdles, a strong state responsibility norm in this context would require extensive and intrusive governmental activity in the private network that at least for now is anathema in the United States and other Western democracies.

A final problem is espionage. *Just and Unjust War* says practically nothing about espionage, perhaps because in both war and peace, international law hardly regulates it. Similarly, international law says little about state-sponsored theft of intellectual property and military secrets. It is unclear whether, in the cyber era, international law's non-regulation of spying and theft can continue. One reason is that the cyber realm makes possible massive theft of intellectual property and military and intelligence secrets that, in the aggregate, can (and many people believe now do) constitute a serious national security problem, a problem that could conceivably require a military response. Another reason is that the software agents that facilitate cyber espionage and those that facilitate cyber attacks are hard if not impossible (*ex ante*) to distinguish. This means that no nation can tell for sure whether the logic bombs and related agents it finds in its civilian infrastructure networks are agents of exploitation or attack – until, of course, they are used for attack. If these agents turn out to be used for attack, our complacency about the agents of exploitation – and about international law's non-regulation of digital spying and digital theft – will surely change.

Cyber and international agreement

Even if we assume that some of the puzzles in Section 2 can be worked out, and that nations of the world can agree in theory on how *jus ad bellum* and *jus in bello* should apply to the cyber realm, a further hurdle stands in the way of developing true international norms to reflect this substantive agreement (Goldsmith 2010). The main hurdle is verification, which is difficult in the cyber realm because attribution is challenging.

If one sees the laws of war in instrumental terms, akin to solutions to a prisoners' dilemma, then it is doubtful that a rational government would forego using otherwise desired cyber exploitation or cyber attack capabilities in compliance with an international norm in exchange for mutual restraint by adversaries. The main reason is that a government cannot tell whether its adversary is complying with the norm, and thus it cannot know (until it is too late) whether it is receiving a benefit for its restraint. Moreover, cooperation in the prisoners' dilemma depends on credible retaliation when there is breach. Uncertainty in attribution makes retaliation for breach much harder for any president or general to order. ('Sir, we are 28 percent sure the Chinese did it.') And this in turn makes retaliation less credible to some probably large degree, which in turn invites breach and unravels cooperation.

But perhaps the laws of war work more through normative influence. Once the norms are established and accepted, they will exert normative pressure on states to comply. I think this can work in institutions like the US Department of Defense, which have massive bureaucracies of relatively independent lawyers and compliance officers devoted to following the laws of war and a strong culture of compliance. (These bureaucratic structures likely have an instrumental foundation, but I will set that aside for the moment.) And, in fact, there have been many reports that the DoD is deeply self-constrained – some say too constrained – by the laws of war in its use of offensive cyber weapons. But there are few militaries in the world with the type of self-constraining bureaucracy as the DoD. Can the laws of war have a normative influence on compliance with these other countries?

I am skeptical. There are many reasons for skepticism, but the main one is the problem of attribution. The laws of war would not be nearly as efficacious, or have the same level of normative salience, if the nations that violated the laws could not be identified and publicly shamed. Norms cannot get much purchase in a world without serious attribution; anonymity is a norm destroyer. That, unfortunately, is the situation in the cyber realm. The problem is exacerbated by the fact that, even if a nation has perfect attribution, it often cannot publicly reveal the evidence of attribution because doing so would disclose espionage and attribution capabilities and render them less useful. To the extent that this is so, it makes the public-shaming aspects of a verification regime, and thus the operation of norms, less robust.

In this world of anonymity, it is unlikely for the laws of war to have much normative purchase to constrain nations that lack robust bureaucratic commitments to compliance with the laws of war and that are otherwise inclined to use cyber weapons. Stewart Baker imagines the differences between the DoD and most other military bureaucracies like this:

> The Pentagon would be exquisitely sensitive to arguable violations of international law in carrying out operations in cyberspace. Our guys would sit with their fingers poised over the 'return' button for hours while the JAGs were trying to figure out whether the Belarusian remarks in committee were a consensus or an individual interpretation of article 42bis. And nobody else would give a damn what the treaty said, because they wouldn't expect to get caught and because even implausible deniability can't be rebutted with the certainty needed to make a legal case, let alone send missiles in response (Baker 2010a, 2010b).

Baker is exaggerating for effect, but his essential insight about the relatively robust compliance commitment by the United States and its allies as compared to other nations, and the resultant opportunities for mischief and opportunism in the cyber realm as a result, is right.

The many hurdles to developing international norms do not necessarily mean that the growing stockpiles of cyber weapons will lead to cyber war. Above I noted that norms would not have much influence on nations '*otherwise inclined to use cyber weapons*.' The moderately good news, I think, is that, although many nations have the capabilities to engage in large-scale cyber war, they are not using the weapons because doing so would be self-defeating. Even with the cloak of relative anonymity, the potential catastrophic costs to the globally integrated computer and telecommunications infrastructure from a large-scale cyber war creates powerful disincentives for a nation to engage in such war. One can, if one likes, call mutual restraint of this sort a 'norm.' This norm is nothing more than a behavioral regularity resulting from the coincidence of uncoordinated self-interest among nations. But if every nation continues to have such independent incentives, cyber war might not emerge.

To see the point, consider Richard Clarke's claim in his book *Cyber War* that China will in the near future engage in cyber attacks on the United States (Clarke and Knake 2010). It is true that China has significant offensive cyber capacities that could in theory cause enormous destruction, and that it is stockpiling cyber weapons and planning for cyber war. But the same is true of the United States. What Clarke never adequately explains is why China or other nations would use these weapons in this way. Capacities and contingency plans, taken alone, do not add up to a serious threat. There must also be a plausible scenario in which a nation has the motivation to use these weapons.

Clarke addresses this issue briefly in trying to explain why China might destroy American infrastructure by means of a cyber attack even though 'China's dependence on US markets for its manufactured goods and the trillions the country has invested in US treasury bills mean that China would have a lot to lose' (Clarke and Knake 2010). He says that the United States and China might nonetheless be drawn into a war over Taiwan or the oil-rich islands in the South China Sea. Perhaps, but it is hard to imagine that China would wipe out the New York Stock Exchange or the electrical grid of the East Coast unless it were in a total war over those islands – the sort of war that would also involve enormously destructive non-cyber weapons. Clarke is also right that China's cyber weapons might (like China's conventional forces) deter the United States from intervening against China in a Pacific Rim contest. But he should also acknowledge that this deterrent is weakened by China's dependency on a functioning American economy, which significantly reduces the credibility of its cyber threat.

I am not saying that there is no chance that a nation might want to use cyber weapons for attack, possibly rising to the level of war. We have already seen low-level cyber attacks related to war in Georgia and Lithuania. Many people think that Stuxnet is the first truly dangerous cyber weapon and that it was designed by the Israelis to knock out the Iranian nuclear weapons program. It is also possible that the stealth cyber-arms race, the difficulty of knowing for sure which nation is behind a cyber attack, and the general absence of effective norms to govern such attacks combine to create an unstable situation in which destructive cyber activities might escalate by accident. Finally, criminal groups have growing capabilities that could cause significant damage to nation-states, and terrorists are now in the market for these capabilities. All of these developments are worth worrying about, and will present enormous challenges to, and likely require large changes in, our understanding of the laws of war.

Note

1 This chapter has been published previously in the *European Journal of International Law*, 24(1): 129–38.

Section II

Geopolitics of Conflicts in CyberSpace

5
Russia's Information Warfare Capabilities

Roland Heickero

Introduction

One way of understanding Russia's actions in the information arena and how it views threats and the protection of Russian sovereignty is to study officially published doctrines, policy documents, and statements by representatives of the regime. Since the fall of the Soviet Union, two military doctrines have been published, as well as several white papers, in response to various political and military events. The first doctrine was launched in 2000 (Sokov 2007). From it we can gather that three kinds of conflicts, above all, are seen as threats to the nation. The first threat involves the risk that conflicts escalate in the immediate vicinity of the Russian border. The second is the risk of a direct military confrontation with the United States and its Western allies. The third involves a conflict with an expansive China.

In the aftermath of the Georgian war of 2008, President Vladimir Putin ordered an overhaul of the doctrine from 2000, since it, to some extent, was no longer up to date. In December 2008, plans were announced for a new Russian doctrine, which was subsequently published in 2010 (The Russian Government 2010). The doctrine contains a change from the previous version. The rhetoric against the West, and specifically NATO, was amplified. This increased focus on criticism of the West has been interpreted as the military wanting to reinforce its position. One reason for this is that Russia sees NATO as expanding at its expense and that the nation is about to be surrounded by countries that are viewed as opponents.

The Russian attitude is in many respects influenced by historical experiences like the Great Northern War of 1700–21 with the eastern campaign of Charles XII of Sweden; the march of Napoleon's *grande armee* toward Moscow in 1812, also known as the Russian campaign; World War I and II; and, of course, the Cold War. Added to this, there are a number of other conflicts in which the country has been involved. The basic view is that military conflicts should, as far as possible, be avoided on Russian soil. Protection of the nation should be performed by using buffer states, which includes the *near abroad*.

President Vladimir Putin and people linked to the administration have, in various contexts, expressed the regime's view in terms of 'protection of Russian interests abroad.' This includes Russian-speaking populations that can be found in areas bordering Russia, such as the Baltic States, Ukraine, Moldova, and others. However, the protection offered is more comprehensive than that. On 6 March 2014, Prime Minister Medvedev, via the ITAR-TASS news agency, announced that it might be possible that Russian citizenship will extend to Russian speakers in territories that previously belonged to the Russian empire. If so, this would include countries such as Finland, which under very severe conditions and civil war between *red* (socialist) and *white* (conservative) groups, broke away from the Russian hegemony in 1918. Russian soldiers based in Finland participated on the *red* side in the conflict.

In order to create a momentum for expanded courses of action in its areas of interest, and to protect the Russian population, Russian passports have repeatedly been distributed among ethnic Russians in border areas. This has been done in the Baltic States, East Ossetia, and Ukraine, among other places. According to one report, up to 40,000 Russian passports have been distributed to ethnic Russians in Ukraine. There are suspicions that people with a background in the security service, the Federal Security Service (FSB), and the Foreign Intelligence Service (SVR) have been involved. Naturally, this causes great concern in the countries that are targeted. For example, the Estonian security service has identified that there are what is described as *illegals* in the country, with probable links to the Russian security services.

The use of *illegal* in this specific case originates from the Soviet Union and is also used to indicate intelligence officers who infiltrate a country a long time in advance, without diplomatic immunity, with false papers and a false life story. The purpose is to prepare for specific intelligence operations that may occur at a later date or over time. This technique was successfully used, for example, in connection with the invasion of Czechoslovakia in 1968, known as Operation Danube. More than six months prior to the invasion of Afghanistan at Christmas 1979–80, Russian special units – Spetsnaz[1] – and airborne troops from the Vozdushno-DesantnyeVoyska (VDV) reconnoitered the prospects to occupy the country. Similar methods may have been used in the annexation of Crimea in 2014 (Gyllenhaal and Von Braun 2013).

Apart from illegals, the presence of *agents provocateurs* is another method used to infiltrate a country or regime. In these cases, they are more likely to be local self-defense groups that act more or less independently and arbitrarily, without the express will of the regime. In connection with the unrest in Ukraine, the national security service has designated Russian entities as agents provocateurs with the intent to disturb the order and create chaos. The purpose behind fomenting unrest through the agents provocateurs is to set up conditions that make the case for further Russian military and political actions. Possible links between the agents provocateurs and the Russian security services and the Russian regime are difficult to prove. The accusations may be part of both sides' information warfare in order to influence public opinion.

Projection of power

Protection of Russian sovereignty is created by using economic, military, and political means of pressure or a combination of these. The manipulation of energy is a well-known, potent economic-political means of control and projection of Russian power. Power over vital energy resources can be used as a weapon, particularly when the regime's demands are not met by an adversarial party. For instance, during the worst cold snap in the winter of 2009, natural gas supplies were cut to Ukraine, a transit country for energy to the rest of Europe, affecting 17 EU countries apart from Ukraine.

In 2011, an economic integration project called the Customs Union was formed between Russia, Belarus, and Kazakhstan. Its purpose is to create conditions for free trade in services, labor, and capital between the member countries, similar to the setup of the EU. The plan is for the Customs Union to lead to the Eurasian Union by 2015, incorporating several more regional countries. The aim is both to reinforce relations with nearby states and re-establish Russia's influence globally, thereby tying states closer to the Russian nation in accordance with buffer zones and the near abroad. The Eurasian Union may also serve as an attractive alternative to the EU. The gas agreement between Russia and China of May 2014, solidifying an energy supply arrangement, can be interpreted as a way to reinforce relations to China but also a way of dividing the West. As a fair amount of Europe's natural gas is supplied by Russia, taking a stand with the US and other allies against Russia may prove to be even more difficult, as Europe will seek to keep energy flowing westward rather than eastward.

Projection of military power has taken place repeatedly since the dissolution of the Soviet Union, most notably in the wars in Chechnya in the 1990s and the 2000s, the Georgian war in 2008, and during the intervention in Ukraine and Crimea in the spring of 2014. Since the Georgian war, Russia has made massive investments in its military capacity. This includes new materiel and equipment such as combat vehicles, improved amphibian capacity with new landing ships, hovercraft, and helicopters, as well as a more efficient command and communications system. Large investments are being made in submersibles and midget submarines adjusted for special operations. Special elite units such as Spetsnaz, the VDV, and naval infantry have been upgraded and rearmed. Some of the special units have electronic and IT warfare capacity and capability for psychological operations. In 2013, it was announced that Russia had established a new, comprehensive organization for special operations outside the national borders: the so-called special operations forces command, or SSO, with its special missions, equipment, and techniques.

In February and March 2013, President Vladimir Putin issued orders for two major exercises to be held specifically for the VDV and the naval infantry in order to test the rapid reaction capability. During the March exercise, they managed to mobilize 7,000 men, 250 armored vehicles, and 36 ships in a couple of hours. For example, the 336th naval infantry brigade, based in the Russian enclave of Kaliningrad on the Baltic Sea, was prepared and ready for the mission in only two hours. The same preparedness time also applies to the guards airborne division stationed in Pskov outside St. Petersburg. These and similar units can carry out coordinated surprise attacks at great operational depth. This sends strong security political signals to the rest of the world that this readiness capacity now exists and that it can be used.

A couple of months later, in September, the *Zapad 2013* (West 2013) military exercise was carried out with Belarus, featuring at least 70,000 men, making it the biggest preparedness exercise since Soviet days. During the exercise, the long-range and mobile Iskander missile system was launched for the first time in the Baltic area (Forss 2012). The system, which is primarily intended for high-value targets such as SAM units, air and naval bases, and command posts, was successfully used during the Georgian war. In December 2013, the semi-ballistic system was deployed as part of the preparedness exercises in Kaliningrad. The system's range is 700 kilometers, which covers the middle and southern parts of Sweden, northern Germany, Poland, the Baltic States, and southern Finland.

In addition to increased military capacity, military rhetoric has also been turned up and has become increasingly confrontational. For example, on 5 June 2012, the Russian Chief of the General Staff, General Nikolai Makarov, warned Finland against aiming for a closer cooperation with NATO.[2] In an interview on 8 June 2014 with Sergei Markov, Putin's personal envoy, he warned that the Russophobia in Sweden, Finland, Poland, and the Baltic states could lead to a third world war (Lauren 2014).

On Good Friday, 29 March 2013, four Russian bombers escorted by two fighter aircraft flew out of the Gulf of Finland, heading for Swedish territory. The aim of the operation was to simulate missile attacks against two of the most important military facilities in Sweden; the one a military air field and the other the headquarters of the national signals intelligence organization, the FRA, outside Stockholm. The cruise missiles were armed with nuclear weapons. Less than a month later, Russian reconnaissance aircraft flew in international air space between the Swedish islands of Gotland and Oland. The aim was to monitor signals communications between staffs in an international communications exercise.

In March 2014, the decision was made to build a new Russian base with a force of 3,000 men and helicopters in Alakurtti, only 50 kilometers from the border with Finnish Lapland and 300 kilometers from Sweden (Jones 2014). A new signals intelligence station is also going to be located in the same place. Its task is to look for threats from the West. The base is also going to cover large parts of the Arctic, an area with increasingly important security, political, and economic interests. The location of the facility has symbolic value; it is situated in the same area as some of the biggest battles of the Winter War of 1939–40 and the continued war in 1941–44 between Finland and the Soviet Union.

At a security policy level, Russia acts resolutely, as in the case of the handling of chemical weapons in Syria. This also applies to the events in Ukraine. The regime often anticipates the inactivity of the international community, which is then forced to react to a *fait accompli*. The agility – when it comes to mobility and speed – of the regime's actions, coupled with calculated risk-taking, can be seen both as an active means of fulfilling its strategic aims and a way of projecting power. This follows the military doctrine of 2010. The strategy has proved successful and fulfilled its purpose.

Information warfare and cyber-security doctrine

In the Russian sphere of government, there are no officially published doctrines or policy documents that specifically describe information operations (IO) and information warfare (IW) in a way that corresponds to the American doctrines or NATO's doctrines on IO. However, there is the Doctrine for Information Security for the Russian Federation of 2000, which deals with different aspects of security in the information arena. The information security doctrine describes threats against the nation and how the Russian state should act in order to protect strategically important information (Carman 2002). The doctrine should be seen as a policy instrument that is, above all, addressed to the Russian society, but it is also intended to influence foreign parties.

The Russian military has claimed that the effect of information warfare can be compared to nuclear weapons. If hostile information warfare is directed against economic or national command and control systems or against the fighting ability of the armed forces, 'Russia reserves the right to resort to nuclear weapons, first against the actual information weapons but later also against the attacking state' (Grau and Thomas 1996).

Information warfare is implemented at strategic, operational, and tactical levels and takes place in peacetime, during preparations before war, and during war (Limno and Krysanov 2003). It covers the whole scale from the political level with ministries down to individual units. Information warfare consists of electronic warfare, psychological operations, intelligence collection, deception, and so-called capacities for *mathematical programming* (Pirumov 1996). The latter can be interpreted as computer and network operations, or conducting of cyber war. It includes both offensive and defensive capability.

The purpose of the Russian cyber warfare capacity, according to a 2001 US Congressional Report (Hildreth 2001), is to gain information superiority. In wartime, this means achieving and maintaining dominance over the enemy in the cyber arena, but also protecting the nation's own crucial information and information systems. During warlike conditions, such operations take place more openly than in peacetime and support conventional forms and methods of warfare (Tsymbal 1995). Information superiority is achieved through the use of specific efforts against the opponent's information systems, decision processes, and command and control systems, just as against the population. Malicious code such as viruses and other information-related weapons make up reinforcement.

In future conflicts, there will be no clear defense and attack lines; the battle will be waged in many dimensions and arenas. This also applies to the cyberspace arena. Warfare continues to shift from duels between weapons systems to duels between information systems (Fitzgerald

1994, 1996). The arms race does not mainly take place in the physical arena but in the digital sphere with software and algorithms as well. The one with the fastest and most advanced computer capacity will have an advantage over the opponent. The main purpose of cyber war, in these terms, is to break down the functional capacity of the opponent's most important military, industrial, and administrative assets and systems. It is about creating an information psychological pressure on the opponent's military and political leadership, its armed forces, and the population (Dylevsky et al. 2007). This is primarily going to be achieved via ultramodern information technology and equipment.

Cyber war means the physical destruction of military information systems, the conducting of electronic warfare, induction of malicious code, carrying out of DDoS attacks and spamming, and using backdoor functionality and logic bombs directed against the enemy's command and control systems. Analysts Kukashkin and Yefimov (1995) have expressed concern about so-called algorithm bombs, or software bombs. The aim of these is to destroy parts of algorithms, thus limiting the functionality of computer programs. That way, erratic behavior can be achieved. Other methods include theft and other forms of illegal collection of information, including cyber espionage, for example. According to Deputy Chief of the General Staff Alexandr Burutin, the information weapon requires

> no special manufacturing equipment or complex infrastructure. A small group, or even an individual expert, can develop and perform a devastating attack and subject people's lives to risks, this without necessarily having to physically cross a border.[3]

A small number of skilled hackers, specialized for the purpose, can create major problems in an opponent's vital information systems. Other information weapons include electromagnetic energy weapons that neutralize electronics and components from a distance, thereby incapacitating them – temporarily or permanently (Rastorguyev 1998).

The official Russian view on cyber warfare

Although there is no specific doctrine for information warfare, Russia is active in the international arena in order to present its view on strategic information security. In the autumn of 2011, it signed a code of conduct known as the *Yekaterinburg convention* with China, Tajikistan, and Uzbekistan, regulating how states should behave in the cyber arena.

Russia is one of the initiators of the UN-led Groups of Governmental Experts (GGE). At an international meeting in Berlin in December 2011 arranged by the German ministry for foreign affairs and the UN disarmament organization UNIDIR,[4] among others, Igor Nikolayevich Dylevskiy, head of the Department at the Ministry of Defense, said that the Russian armed forces are developing principles and rules of conduct for protection of critical systems, in order to protect from, deter, and prevent attacks. According to Dylevsky et al. (2007), the characteristics of the Russian way of acting are openness and strict compliance, global cooperation, and respect for the sovereignty of individual states. The Shanghai agreement was mentioned as an example of openness and the willingness to cooperate. The Russian armed forces have a corps of standing forces with constant preparedness to act 24/7. They can use all measures and precautions to protect critical systems, but with observed caution. In the event of a cyber war, Russia will first consult the United Nations and, subsequently, depending on a possible escalation of the conflict, discuss the matter with colleagues from NATO. The purpose of the discussions is to prevent possible violations of international laws. There is a need for 'norms of international prohibition,' that is, what can be interpreted as rules of engagement (RoE). Russia sees international cooperation between states as something that is important to achieve.

The view of the Russian armed forces can be summed up as them having accepted the challenge from their opponents with the aim of reinforcing national security. Should conflicts arise in cyberspace, the Russian armed forces will act through use of a military deterrent to get the opponent to refrain from cyber attacks via threats. Based on historical experiences and events, a deterrent is almost a reflexive reaction, a first line of defense.

In official contexts, representatives of the Putin regime have said that Russia has the capacity to act globally with all possible means in order to protect its own critical systems. Should conflicts arise in cyberspace, the Russian armed forces will act to deter its opponents. In various official statements from different players, the effect of cyber attacks against vital societal and military systems is compared to weapons of mass destruction, where Russia reserves the right to resort to nuclear weapons against its antagonists if so required.

When it comes to regulations of weapons of information warfare, the Russian armed forces see a need for broad international cooperation in the field, bilaterally established between states and following international laws and rules. However, the question is how events such as the conflict in Ukraine and Crimea will affect the prospects for international agreements in general and more specifically when it comes to development in the cyber arena.

The Russian capacity and capability for information warfare

Since the early 1990s, significant efforts have been made to strengthen the security of the Russian state against both international and domestic threats. Information warfare (IW) is seen as an important capability to develop for both offensive and defensive purposes. The military doctrine from the spring of 2010 stresses the importance of information warfare in the initial phase of a conflict in order to impair the opponent's possibilities for command and control. Throughout the whole conflict phase, targeted and specific information campaigns are pursued in order, in part, to create goodwill in the international community (Vendil Pallin and Westerlund 2010). These campaigns can be used to present a positive view on Russia's way of acting as well as to induce a questioning of the opponent's intentions.

Within the Russian administration, several federal authorities are responsible for handling information-warfare capabilities, including all forms of networked and digital activities. These activities are not limited to the Internet and cyberspace but also cover electromagnetic warfare and campaigns of influence. The military system for collective security is divided according to the regional principle. Each military district has its own capacity for IW. There are four major agencies dealing with IW in a broader sense: the Federal Protection Service (FSO), the Federal Security Service (FSB), the Foreign Intelligence Service (the SVR), and the Military Intelligence Service (GRU). With the exception of the GRU, all these organizations are the result of the breakup of the KGB. The FSO, the FSB, and the SVR are subordinated to the president. The GRU is a part of the Ministry of Defense as the central military intelligence body for the General Staff.

On the strategic level, the FSO has overall responsibility for conducting signals intelligence. The FSB is responsible for internal security, including wiretapping of telephone lines, opening mail, and monitoring other forms of communication such as Internet traffic surveillance. Traffic is monitored by the SORM II system.[5] On request from the FSB, all Internet service providers (ISPs) have to invest in this kind of surveillance system. Actually, more or less all communications transmitted through operators such as Rostelekom, Transtelekom, and Elektrotelekom are forwarded to the FSB (InfoSecurity 2009; Carr 2012). Moreover, the FSB probably has capability for conducting computer network operations including exploitation, attack, and defense.

The Foreign Intelligence Service is also subordinated to the president and is responsible for providing intelligence, operations, and analysis on behalf of the Russian president, the Federal Assembly of the Russian Federation, and the government. Together with the GRU, its main function is to provide the state leadership with foreign intelligence. The SVR conducts human intelligence activities against its adversaries, but it also has the capacity for strategic signals intelligence and managing military and commercial satellite systems and fixed and wireless communications.

The Military Intelligence Service was established in 1918. Over the years, it has changed designation several times. It comprises the foreign intelligence organization of the Defense Ministry and the central body of military intelligence for the General Staff. The GRU is an intelligence system that makes comprehensive use of practically all forces and means of intelligence. It maintains units for signals intelligence, imagery reconnaissance (IMIT), and satellite imagery capabilities (SATINT). It also conducts open-source intelligence (OSINT).

The GRU has over 30,000 staff, divided into 24 individual brigades, each numbering 1,500 men. Under the GRU's command, there are special operational units, or Spetsnaz, responsible for surveillance of communications, electronic intelligence, and psyops (Thomas 2004). In an interview in *Pravda* in 1996, GRU General Feodor Ladygin confirmed that the organization has the capability to hack computer networks in order to collect sensitive information.

Within the Russian Armed Forces, there are also special units such as the signals troops and radio-electronic combat units dealing with electronic warfare capabilities at operational and tactical levels. Sources of signals intelligence (SIGINT) and electronic intelligence (ELINT) are captured by aerial and sea assets as well as by the Strategic Rocket Forces. Another organization connected to the Russian Federation is RU-CERT (the Computer Emergency Response Team), which is responsible for reporting cyber incidents.

Regarding specific IW units for computer network operations, there is very little open information where capability, organization, and structure are concerned. A qualified assumption is that all four intelligence and security services have their own resources for conducting offensive and defensive networked and digital activities due to their specific tasks and areas of responsibility.

Psychological and political control in information warfare

In peacetime and wartime, information warfare is about protection of the Russian state. It is covertly pursued through intelligence collection and through political and psychological moves. It involves diplomacy and various economic tools. At a supranational level, the aim is to influence the views of public opinion and drive wedges in alliances between possible hostile opponents via information operations (Heickero 2010).

As compared to their Western colleagues, Russian analysts attach great importance to control of information and associated psychological impacts in the sense of protecting society against information influence from an opponent (Thomas 2003). The aim for control not only includes state security but also for the regime and leading players' personal interests and ambitions. In the event of a conflict, the regime's view is that the (Russian) media are supposed to present awareness to its audience about what is going on, but it should refrain from exaggerations. The media are supposed to influence to the extent that they should calm the tone of the debate in order to prevent a possible escalation in the public opinion. All this merges in a joint view to protect the nation against possible internal division.

This should be understood in the context of the dissolution of the Soviet Union. Many analysts believe that one of the main reasons for the collapse was hostile psychological operations (Thomas 1998, Hoffman 2008). The fact that the collapse is a stigma that weighs on the current regime is perhaps best described by President Putin's statement that, according to him, the fall of the Soviet Union was the biggest geopolitical disaster of the 20th century.

In accordance with the information security doctrine, stability in society in Russia is reached through control of the media, such as TV, radio, and daily papers. This can be interpreted as the authorities having an interest in regulating information and media networks, for example, through nationalization of free media. The doctrine is a valuable tool for the Kremlin to gain control over the flow of information into and throughout the country. One step in this desire for control is the Putin regime's demands for servers handling social media communications to be moved to Russian soil. Russian operations of influence take place in the whole media spectrum, via traditional channels such as TV, radio, and newspapers as well as on the Internet via YouTube, blogs, and social media.

Through nationalization of media, such as the TV channel NTV and other free channels, the state acquires instruments for monopolization of the truth. The news channel Russia Today (RT), which was started in 2005 and which broadcasts in Arabic, English, Russian, and Spanish to viewers in more than 100 countries, is seen by some as an important tool for the regime to convey its message. This use of the television channel is a key part of information doctrine implementation. The channel has repeatedly been used as the carrier of the government's preferred information and views in connection with conflicts such as the one in Georgia in 2008 and the Crimean crisis in the winter and spring of 2013–14. The purpose is to rouse public opinion in favor of the cause. It is about owning the story, thereby creating the narrative.

They have learned from experience. From a psychological-operational perspective, the Soviet Union failed in its operations of influence against its opponents in Afghanistan (Serookiy 2004). The conflict in Chechnya resulted in the central government's realization of the need to assume control over the flow of information in and from a combat zone and manipulate its psychological effects on society. Both wars in Chechnya show that a small and relatively powerless opponent in some areas can gain information superiority in relation to a stronger party through an efficient use of the media component.

Maskirovka and agents of influence

Maskirovka, that is to say camouflage and deception, is a basic function in Russian information warfare both during peacetime and in times of conflict. It is a means of influencing the enemy by giving him a false sense of control over the situation and thus making him act in a predictable way that is detrimental to his interests. Historically, the Russian military and the political elite have great knowledge of and experience with *maskirovka*. One organization that is responsible for such deception operations is the FSB's Service A Chief Directorate.

One important element in deception at a strategic level as part of information operations and *maskirovka* is the use of *agents of influence*. These are individuals who present views and try to create an opinion around various ideas and subjects for a specified purpose. This purpose could be political, ideological, economic, idealistic, and the like. The opportunity to influence a situation is dependent on the represented individual or organization's credibility with the desired audience. The more credibility the individual has, the more likely he or she is to be able to influence opinion in the desired manner.

Individuals who participate in operations of influence can serve in a number of capacities, such as journalists, in political parties, as authorities, within the arts, at academies, and so on. They may be intellectual leaders, embassy staff, nationalists, religious leaders, and well-known bloggers in social media. Generally speaking, agents of influence fall into three categories based on an individual's level of involvement and motivation. The first category is agents that have been directly recruited by a foreign security service or the like. They can be stationed in a specific environment and be controlled by the client. The second category is a reliable contact, who is not controlled or recruited by a foreign power in the same way but who is cooperative. The third category, finally, is people who act as useful idiots without understanding that they are playing someone else's game.

It is very challenging to prove that an individual or an organization is an agent of influence and is being managed by a foreign power or security organization. During the Cold War, there were suspicions in the West that native people at very senior political levels had been recruited for such purposes by the Soviet Union and the German Democratic Republic. However, no cases against agents of influence were ever pursued or proven.

Using agents of influence to undermine and question in order to present one side's views and influence the story in a subtle way is a prominent technique of information warfare. For the present Russian regime, this way of acting is interesting, especially in view of the difficulty in linking this manipulation to a potential instigator and underlying interested party.

Propaganda and other psychological information warfare operations used during conflicts

The annexation of Crimea and the subsequent incidents in the eastern parts of Ukraine clearly show the approach and realization of psyops in connection with conflicts. The text has previously mentioned that so-called illegals and agents provocateurs can be used to create a momentum. To this, one can add agents of influence, who present views that favor the own side, and alternatively can create confusion and a questioning of the opposing side's actions.

Examples of psychological operations and information campaigns in the wake of the Crimean crisis have been described by journalist Timothy Snyder (2014) in *The New York Review of Books* and by historian and Pulitzer Prize winner Anne Applebaum (2014) in the British newspaper

The Telegraph. Among other things, Applebaum mentions how Western media have uncritically spread the information that 675,000 Ukrainians have fled to Russia. This was supported by photos of lines of cars at the Ukrainian border. However, it soon turned out that the photos were old and from the daily lines at the border crossing to Poland. Other psyops rumors have included information that the firm Blackwater had landed mercenaries in Ukraine and that neo-Nazis had taken over the Ukrainian parliament in Kiev. Neither was true.

The target of the false reports and propaganda in these cases is not primarily the West but Russians and Russian speakers. English speakers use the state news corporation Russia Today, which in addition to communicating in English, broadcasts in a number of other languages. Today the TV channel has attained a reliable status and is often quoted in Western media, with comments functions and news anchors resembling those in the West. This Western-style presentation is part of the battle for public opinion.

In the cyber operations that were initiated by nationalist hacktivists against Estonia in 2007, in connection with the relocation of a Soviet military monument in the capital Tallinn, and the five-day war against Georgia in 2008, there were elements of psychological operations. Apart from massive DDoS attacks and interference with communications systems, there were prominent visual reminders of information command and control. For example, there were defacements of Georgian governmental websites with scurrilous portraits of President Shakashvili looking like Adolf Hitler.

During the annexation of Crimea and the incidents in the eastern part of Ukraine, extensive psyops took place on pro-defense blogs.[6] According to Dr. John Schindler, professor of national security affairs at the US Naval War College, people who discuss the Russian propaganda in a critical way have been subject to vicious attacks on the Internet.[7] In addition, spam bots have allegedly approached several leading defense bloggers, and comments and discussion forums have been filled by pro-Russian propaganda. Specified content messages are automated and notices are distributed and propagated to social networks en masse and in several languages. There are suspicions that translation programs are being used to craft these messages and notices, since word order and grammar have not always correlated. According to unconfirmed information, the Foreign Intelligence Service has developed this capacity. Apart from being contractually employed at the security service, the instigators can be made up of nationalists who utilize various methods, such as hacktivism, to act out their own agenda. At an official and tactical level, during the annexation of Crimea, the 22nd Spetsnaz guards brigade is believed to have been involved with its psyops unit.

The case of Edward Snowden

Edward Snowden makes up a special chapter in Russian operations of influence. In 2013, Snowden was runner-up in *Time* magazine's Person of the Year poll, demonstrating his place as one of the most influential people in the world due to his revelations about the National Security Agency's surveillance apparatus. This does not mean that he is an agent of influence for Moscow *per se*, though Snowden does fill a role in the Putin regime's agenda and narrative against the West and more precisely the United States' spying capabilities. The reasons for his actions are multifaceted, and his motives have likely changed over time. Both the timing and the locations, first to China and then to Russia, give rise to the question about the benefits of his defection.

At the time of his defection, there was an intense discussion in the media and in diplomatic talks between the Obama administration and its Chinese counterpart about alleged Chinese cyber espionage against American targets. Snowden's revelations about the NSA's mass monitoring effectively put a stop to the discussions for the foreseeable future. For a period, the United States has been forced to be reactive on issues involving strategic information security, especially versus its cooperation partners. US credibility has been questioned in many contexts, and these questions are subsequently used in the Russian side's information operations. Apart from a necessary and important discussion about integrity and the rule of law in an environment of mass monitoring in the wake of the NSA debate, Snowden's revelations have also had economic effects for American IT companies.

According to a report from the think tank Information Technology & Innovation Foundation, American IT companies may lose up to $35 billion in revenues up until the end of 2016 (Gaouette 2013). While this is happening, competitors are advancing their positions on the market. In addition to lost sales, the report warns that the NSA debate may also lead to the obstruction of US attempts to promote an open Internet, thereby affecting large American information technology companies that depend on an open web to reach their clients worldwide. There are a number of countries, including the US, that are looking to reinforce their national control of both the networks and the information that is accessed and disseminated, something that China and Russia and other states have already done. In the long run, there are risks that the

Internet will be divided up into regional structures so as to provide a better mechanism for control of information.

From a strategic perspective on information security, the NSA debate may lead to a reinforcement of the work on increased openness on the Web. However, there is also a risk that the effect will be the opposite, which would have unfortunate economic and social consequences. Just as the NSA's activities are now being examined, the activities in cyberspace of the Russian and Chinese intelligence and security services should also be questioned as there are clear similarities between the US signals intelligence capacity and its Russian and Chinese intelligence and security services counterparts.

Sixth-generation warfare

Examining the annexation of Crimea, the five-days war with Georgia, and the hacktivist attacks against Estonia from an operational point of view is useful in order to understand the Russian actions in both the physical arena and in the cyber sphere. These events give new insights into Russian military conduct and capability when it comes to information warfare and psyops, including deception and *maskirovka*. This is clearly a new approach in line with what some analysts designate the sixth generation of warfare.

During the annexation of Crimea, the Russians succeeded in combining political and diplomatic pressure, economic sanctions against Ukraine, organization of political opposition, and other non-military means with the use of military force. The operation in Crimea utilized many different techniques, such as propaganda on the Internet and via traditional media; psychological and economic warfare against the target; deception through camouflage of Russian military personnel where grades and nationality insignia were removed; sound and light discipline with radio silence in order to prevent enemy signals intelligence; and surprise attacks in order to create confusion, gain time, and reach information superiority. Initial success in both the physical arena and cyber domain was a decisive factor in Russian conduct in order to both proceed with and preserve the initiative (Berzins 2014).

The operation against Estonia is one of the first official cyber attacks directed against a country, initiated by nationalist civilians, and carried out with large networks of compromised computers used for DDoS attacks. In the Georgian case, the methods were even more refined, and the cyber conflict took place in close coordination with a corresponding military offensive. Both Estonian and Georgian offensive cyber operations were well coordinated, and those behind these attacks had seemingly good knowledge of which websites to attack and how to block them. Prior to the occurrence of these operations, Russia investigated their opponent's networks and computers in order to identify vulnerabilities.

The cyber operation against Georgia was carried out in a similar way to the one used against Estonia (Heickero 2013). Those who acted were nationalist civilians and groups, possibly with assistance from cyber criminal organizations such as the Russian Business Network (RBN). Social networks were one of the main tools for recruitment of potential hacktivists and as the location where they could put malicious code in the hands of the hackers. Three methods were mainly used: DDoS, SQL injections, and defacements of websites. All three are fairly unsophisticated, but they were applied in an innovative way.

The targets were government agencies and news websites as well as financial institutions. The attacks decreased Georgia's capacity to defend itself against the Russian invasion; the defenders were forced to divide their resources between different activities and areas. Apart from this, there was a psychological handicap; the government's means of communicating with its citizens were impaired. What looks like a coordination of the cyber campaign and the military offensive is probably not a coincidence. However, any links to Russian authorities are very hard to prove. The Russian government denies all accusations about complicity in the cyber operations, nor is there any evidence that it initiated or carried out the operation.

These events show a new course of action that may become normative for future cyber conflicts. In theory, it is possible for a player to use nationalist hackers while denying any participation. They can benefit from the effects at the strategic level without having to face any of the risks. Apart from this, the cyber weapons can be used to create psychological pressure on the opponent in order to get him to act in a way that is detrimental to him.

The effect of the cyber operation against the Georgian communications and information infrastructure was limited, due to the fact that Georgia's infrastructure was poor and Internet maturity in the country was low. Irrespective of this, the case shows what consequences may arise. A well-organized digital campaign against a more advanced country with a large dependence on functioning systems and networks can give considerable effects. This was shown by the Estonian case.

Conclusion

Since the fall of the Soviet Union, Russia has developed new operational capabilities in both the physical arena and the cyber sphere.

It is significant that information warfare is becoming an increasingly important and, in some cases, decisive component in conflicts. The capacities include, among other things, electronic warfare, psychological operations, intelligence collection, computer and network operations (cyber warfare), and *maskirovka* (deception and camouflage). This includes both offensive and defensive capacities. Russian information warfare is an ongoing process that takes place in peacetime, during preparations for war, and during war. It encompasses the whole political scale from the ministries down to individual units. It is about protecting the state during peacetime and has more offensive aims during conflict. Information warfare is carried out in secret, through intelligence collection and via political and psychological operations. It involves diplomacy, strategic communication, and various economic tools. Effective cyber war also features information and psychological processes, including propaganda and agents of influence. Through these information operations, public opinion can be influenced and wedges driven into alliances between possible enemy opponents (Donskov and Nikitin 2005).

Russia has a long tradition of advanced military thinking. They are continually testing new concepts and methods at strategic, tactical, and operational levels. Experiences from the annexation of Crimea in 2014 with advanced deception and surprise attacks, the combined and coordinated military operation with hacktivist cyber attacks in the Georgian war in 2008, and the cyber operation against Estonia by hacktivists in 2007, have given rise to new insights and modus operandi. Russian military and political conduct is in line with what some analysts call the sixth generation of warfare.

Control of the media component has become an increasingly important instrument in the regime's toolbox for information warfare. This applies to traditional media such as radio, TV, and newspapers, as well as the Web with blogs and social media. During conflicts, Russian TV companies are used as a means of broadcasting the regime's message. In order to create confidence and recognition among the public, media outlets are built up and presented in ways similar to their Western counterparts. In various popular blogs, there is more or less obvious pro-Russian propaganda that appears coordinated in time and space with current events. Conscious media logic like this is well in line with the Russian doctrine on information security. The aim is to take control over the story and its narrative, thus creating information superiority over its opponents. While this is taking place, the regime is working for increased inner control of the public's access to information, specifically through nationalization of TV channels and demands for servers that handle social media to be moved to Russian soil.

As outlined in this chapter, Russia is developing a wide range of capabilities for information warfare. Conceptually and method-wise, the Russian military is at the front edge when it comes to information warfare. They act within a number of arenas and domains at the global, national, and regional levels in both the physical sphere as well as cyberspace. Combinations of capabilities are continuously being tested, and their expertise increases over time. Some examples of this are (a) fast and surprising military operations coupled with psyops, deception, and hacktivism on the Web, (b) projection of power through diplomatic and economic pressure via coordinated strategic communication delivered in different kinds of information channels and with different methods of influence, (c) intelligence collection via signals intelligence, cyber espionage, and mass surveillance of traffic, to name a few.

Notes

1 *Spetsnaz* is a joint name for units with special tasks and capabilities. Such units exist in all defense branches, such as the military intelligence service, the GRU, the security service, the FSB, the airborne troops, the VDV, the naval infantry, and the border troops.

2 Russian general warns Finland about NATO. YLE Nyheter Inrikes. Svenska.yle .fi/artikel/2012/06/05.

3 Speech in Info-Forum, 10 February 2008, referred to by Carr in AppSec Asia Conference, 17 November 2009.

4 Challenges in Cyber Security: Risks, Strategies and Confidence-Building. International Conference in Berlin, December 13–14, 2011. Auswärtiges Amt, Freie Universität Berlin, IFSH, UNIDIR.

5 SORM is a Russian acronym for System for Operational-Investigative Activities. It could be compared to the FBI's Carnivore and the British Government Technical Assistance Centre (GTAC); see Leijonhielm, J., Hedenskog, J., Knoph, J., Oldberg, I., Unge, W., Vendil, C. (2000) 'Rysk militär förmåga i ett tioårsperpektiv. En förnyad bedömning 2000.' Användarrapport FOA-R-01758-17--SE (Stockholm, FOA). [Russian military capability in a ten-year perspective.]

6 Some examples of blogs discussing Russian psyops include Cornucopia (cornucopia.cornubot.se) and Wiseman's Wisdoms (wisemanswisdoms.blog spot.com).

7 John Schindler. Mobile.twitter.com/20committee/status/446472439313100800

6
The Sino-US Digital Relationship and International Cyber Security[1]

Jyh-An Lee

Introduction

President Barack Obama and Chinese President Xi Jinping had a two-day summit at Sunnylands in Rancho Mirage, California, in June 2013 (Campbell 2013). This meeting was the first of two since Xi had assumed the presidency in March (Pace 2013). Some Sino-American relations experts agree with Joseph S. Nye Jr.'s statement that '[t]his was the most important meeting between an American president and a Chinese leader in 40 years, since Nixon and Mao' (Senger 2013). Indeed, Xi likened his meeting with Obama to Nixon's historic visit to Beijing in 1972 (Madhani 2013). The Chinese media confidently declared that the 'milestone' meeting would '[put] forward a glimpse of what China's future might look like when it catches up with the US' (Global Times 2013). While the meeting yielded some clear policy breakthroughs on North Korea and climate change issues, the two leaders did not reach a consensus on cyber-security problems, a major irritant between the world's top two economic powers (BBC News 2013). Cyber security has certainly become one of the most important and difficult issues to surface in this bilateral relationship.

According to White House National Security Adviser, Tom Donilon, Obama told Xi that 'if [the cyber-security issue is] not addressed...[it is] going to be a very difficult problem in the economic relationship and [is] going to be an inhibitor to the relationship really reaching its full potential' (Madhani 2013). Obama further offered some examples and articulated that the US government is certain that a number of Internet break-ins were from China. In fact, China's cyber attacks are not a new concern in either the United States or the international community (Jensen 2012; Preston 2012; Goldsmith 2013). The US government has long been aware that China wields formidable capabilities in the area of information warfare (Geers 2011; Poindexter 2013; Steinnon 2010; Jense 2002). The Obama administration had previously called for China's collaboration in setting new rules for international cyber security (Donilon 2013; Etzioni 2013). When Xi took office on 14 March 2013, Obama called Xi to congratulate him and took the opportunity to address US concerns about cyber security (Holland 2013). These instances exemplify the urgency characterizing US expectations of a positive response by the Xi administration to cyber-security issues.

In recent years, cyber attacks emanating from China have also troubled Western companies suffering from significant economic losses due to online intellectual property theft (Hagestad 2012; Schmidt and Cohen 2013; Bambauer 2014; Nakashima 2013). According to reports, some American intelligence officials suspect that the hacking and stealing of American intellectual property has become constitutive of an important strategy by which China hopes to maintain its high economic growth rate. In short, China's cyber capability appears to pose a serious threat to the American economy. Because Chinese hackers have posed a serious threat to both the public and private sectors in the United States, the US government has prioritized this issue in dialogues with China.

China has become an important focus of Internet law literature in recent years (Benkler 2006; Goldsmith and Wu 2006; Lessig 2006; Murray 2010; Zittrain 2008). Most literature concentrates on the nation's strict domestic Internet regulations and technological control of online speech and the flow of information. However, less attention has been drawn to China's aggressive and systematic hacking into the global Internet network. As every society becomes increasingly reliant on digital networks, current and potential vulnerabilities to China's cyber hostilities deserve more attention from Internet law scholarship. This essay will first illustrate the dangers that these China-based threats pose to the world's governments and enterprises, especially American ones. Then, it will analyze several proposals to resolve or minimize the international cyber-security problem and their feasibility. This essay's focus will rest on the positions of the United States and China because the two powers likely to have the most complicated interests at stake and the greatest influence on the issue overall.

The threat of cyber attacks from China

China is not the only country suspected of potential involvement in cyber attacks. The United States is widely believed to have developed cyber weapons against other countries, such as Iran (Chander 2012; Land 2013; Lee 2012; Lee and Liu 2012; Ling 2011; Hathaway et al. 2012). Evidence shows that countries such as India, Israel, Iran, Germany, North Korea, Pakistan, Russia, and the United Kingdom are all involved in cyber hostilities to varying degrees (Hjortdal 2011; Moore 2014; Shackelford 2013). Nevertheless, China may head the threat list. As a point in fact, Obama pressed Xi on cyber security at the Rancho Mirage summit because China is believed to have the most aggressive cyber-intrusion capability in the world and cyber attack has become 'America's No. 1 security threat,' as James Clapper, Director of National Intelligence, pointed out (Chang 2013; Steadman 2013). In addition to the United States, countries such as the United Kingdom, Germany, France, India, New Zealand, and Taiwan have reported suffering compromised data due to cyber attacks originating from China.

Reports suggest that the Chinese military has exploited vulnerabilities in critical components of the US infrastructure for use in any future conflict requiring cyber attacks on the United States. A cyber attack traced to China on an American nuclear arms laboratory in 2007 might have already resulted in the transfer of American nuclear weapons secrets. Other cyber attacks emanating from China include at least intrusions into the American electricity network and airbases. Security experts have warned that Chinese hackers may paralyze American cities by attacking their electricity grid or break into the computer systems of other highly sensitive plants (Rid 2012).

In addition to cyber attacks targeting the US government, in recent years an increasing number of American firms, including Apple, Twitter, Facebook, *The New York Times*, *The Wall Street Journal*, and the *Washington Post*, have been attacked by Chinese hackers. American cyber-security firm Mandiant reports that it has traced a number of network intrusions from a Chinese military unit, while another firm, Kaspersky Lab, has linked a Chinese-based hacker group to attacks involving more than 350 victims in 40 countries. The cyber-security company McAfee has openly asserted that a 'state actor,' very likely to be China, may be participating in a wide range of cyber-attack programs whose targets include the United Nations, American companies, and various governments. Other evidence suggests that a host of notable cyber attacks derive from the People's Liberation Army Unit 63198 in China (Teplinsky 2013, Eichensehr 2015). A notable example of one strain of these cyber attacks took place from 2009 to 2010, when Google and a number of US companies experienced online breaches (Choucri and Goldsmith 2012; Ferraro 2014; Lee et al. 2013; Lobel 2012). Both Google and the

US government have asserted that the Chinese government initiated the attacks in order to hack the Gmail accounts of Chinese human rights activists. This event led to recriminations between the US and Chinese governments, and eventually caused Google to close down its search engine in China. Thereafter, in 2011, Google again suffered similar hacks that were traced to China and that were targeting Gmail accounts.

The tension between the two world powers over cyber security was ratcheted up in May 2014 when the United States accused five Chinese military officers of hacking into American nuclear, metal, and solar companies to steal trade secrets (Barrett and Gorman 2014; Riley et al. 2014; Parajon 2014). The indictment was the first criminal hacking charge against foreign officers in the United States (Ackerman and Kaiman 2014). According to the indictment, the hacking was from Unit 61398 of the People's Liberation Army. China unsurprisingly denied all these charges and described the allegations as 'extremely ridiculous.' Consequently, Chinese officials announced they would suspend participation in the US-China cyber working group. This is another case which US officials may use to distinguish themselves from their Chinese counterpart: the US officials asserted that they do not steal secrets to give advantage to US companies, but in China, the government's line between military and commercial-purpose hacking is unclear (Kaplan 2014).

According to a National Intelligence Estimate, 'which represents the consensus view of the US intelligence community,' the United States is now a target for China's 'massive, sustained cyber-espionage campaign.' It is also reported that according to estimates by American officials, US companies lose $250 billion each year to online intellectual property theft, most of which is attributed to Chinese hackers. Commenting on these estimates, Keith Alexander, director of the National Security Agency (NSA) and head of the US Cyber Command, described the cyber attacks on US enterprises as 'the greatest transfer of wealth in history.' The most serious and grim warning was recently made by Richard Clark, a former cyber-security and cyber-terrorism advisor for the White House:

> Every major company in the United States has already been penetrated by China [...] My greatest fear [...] is that, rather than having a cyber-Pearl Harbor event, we will instead have this death of a thousand cuts. Where we lose our competitiveness by having all of our research and development stolen by the Chinese. And we never really see the single event that makes us do something about it. That it's always just below our pain threshold. That company after company in the United States spends millions, hundreds of millions, in some

> cases billions of dollars on R&D and that information goes free to China [...] After a while you can't compete (Rosenbaum 2012).

If the above charges against the Chinese government are true, China has transformed the international practice of cyber espionage. In the past, government intelligence would break into the networks of foreign governments for political, diplomatic, or military purposes, but they did not hack into private companies to steal intellectual property for commercial interests or economic gains. But according to a National Intelligence Estimate, few countries have 'engaged in hacking for economic intelligence,' and the offense by China is the most serious one, which has become a direct threat to the American economy. Some believe that the private and commercial information stolen by Chinese hackers include 'large volumes of intellectual property [such as] technology blueprints, proprietary manufacturing processes, test results, business plans, pricing documents, partnership agreements, and emails and contact lists from victim organizations' leadership' (Mandiant 2013). In this sense, the aforementioned alarmist talk of Alexander and Clarke appears to be anything but an exaggeration. This may also explain why United States officials have endeavored to draw a line between governments' spying for intelligence purposes and for commercial purposes (Lotrionte 2015). The baseline is that the latter should never be permitted.

There has been extensive evidence that most of the cyber attacks from China are government sanctioned. Although it is not clear whether the hackers in some cyber attacks emanating from China were called into action by the Chinese government, they have shown a certain degree of obedience to the Chinese government in their hacking activities. The Chinese government has never admitted to involvement in any of these attacks. One exception is the Chinese government's one-time public admission that it had attacked the Falun Gong website whose servers were in Alabama. The government appears to have responded to every other accusation of cyber espionage with bold denials (Barboza 2013). So has the Xi administration. During Obama and Xi's summit in 2013, China State Councilor Yang Jiechi described China as a victim, rather than an initiator, of cyber attacks:

> China itself is also a victim of cyber attacks, and we are a staunch supporter of cyber security...on cyber security, China and the United States both are faced with similar challenges. Cyber security should not become the root cause of mutual suspicion and friction between

> our two countries. Rather, it should be a new bright spot in our cooperation (Madhani 2013).

One interpretation of Yang's statement might be that it reflects a cliché: the Chinese government never admits to its underhanded practices. Sometimes, in fact, the Chinese government denies true accusations publicly. Another interpretation of Yang's words is that they are an implicit condemnation of similar US hacking targeting Chinese computer systems; therefore, 'China and the United States both are faced with similar challenges' (Madhani 2013). Such an inference may rest on a murky recent history of American and Chinese hackers engaged in fierce digital combat against important targets in each other's respective countries, and China's development of cyber weapons and hacking capabilities is due to pressures caused by American technological power. China's claims seem plausible after Edward Snowden disclosed the fact that the US conducted 231 offensive cyber operations against China, Russia, Iran, and North Korea in 2011 (Gellman and Nakashima 2013). Before the Rancho Mirage meeting, when China described itself as a victim of hack attacks, it occasionally suggested that the attacks could be traced to the United States (Jones 2013a). It is certainly the case that some American researchers have urged their government to strengthen the destructiveness of its cyber weapons in the struggle with China. As Goldsmith (2013) pointed out, 'It is true that China has significant offensive cyber capacities that could in theory cause enormous destruction, and that it is stockpiling cyber weapons and planning for cyber war. But the same is true of the United States.'

A more optimistic explanation of Yang's statement is that China is signaling its preparedness to work with the United States bilaterally in resolving this issue, which might even constitute 'a new bright spot [for] cooperation.' Xi also sanguinely asserted that through good-faith negotiations, this issue is 'a positive area of cooperation' for the two great powers. This optimistic viewpoint has recently found expression in a statement made by new Chinese Premier Li Keqiang on 17 March 2013: 'I think we should not make groundless accusations against each other, and [we should] spend more time doing practical things that will contribute to cyber-security' (Jones 2013b). Based on Xi's and Li's words, one may have a much higher expectation from the Xi administration than from Xi's predecessors that China will cooperate with the United States and other Western countries in setting international rules for cyber security.

Although people may have different views on China's public assertions, China's real intent remains to be seen. A more practical approach to addressing this complicated issue is to evaluate how feasible various alternatives to dealing with cyber-security mistrust are for China, the United States, and the international community. Therefore, this essay will evaluate several possible alternatives from a legal and institutional perspective.

Solutions to international cyber attacks

Recent cyber-law scholarship and practice indicate that the virtual world can be effectively regulated because physical facilities enabling all online activities are still subject to government control. This statement may be partly true insofar as it concerns only the legal relationship between governments and the private sector. What the statement fails to address is the type of scenario where a national government's online action may negatively affect another national government or citizens in other countries. As one national government does not have sovereignty over another, the latter's behavior will not be subject to the former's regulations. Therefore, without an international enforcement mechanism, what a nation can do in response to another nation's aggressive behavior is quite limited. This is not only a quandary regarding international cyber-security dangers posed by China and other digital powers, but also a fundamental problem in international law. Facing the Chinese cyber threat, the Obama administration has exerted its executive power to collect intelligence from industries about network intrusions (Weissbrodt 2013). Public and private institutions, whether singly or in concert with each other, have developed various technologies to secure digital networks. The government has tried to curb China's hacking activities through a series of indictment and criminal efforts (Lotrionte 2015; Parajon 2014). However, such efforts are apparently inadequate for satisfactorily resolving the problem at hand. Commentators have suggested that the United States take political and diplomatic approaches, such as expulsion of diplomatic personnel, and the imposition of travel and visa restrictions. However, the US government should carefully assess the political costs of these decisions.

As cyber security and cyber war are quite new concepts in international law, the legal definition of cyber war and its relationship with state violence are yet unclear. Nonetheless, as cyber security has been elevated to a national and international security issue, the international community will continue to desire collaboration between the world's great powers in developing and maintaining a workable order. The desirability of such an order grows even more obvious when one considers that the United States, despite its significant investment in cyber-security technologies, cannot repel hacking activities. Therefore, it should come as no surprise that during his meeting with Xi, President Obama told the media that both he and Xi 'recognize that the issue of cyber security and the need for rules and common approaches to cyber security are going to be increasingly important as part of bilateral relationships and multilateral relationships' (Madhani 2013). Below, this essay will discuss some approaches to such collaboration from a legal and institutional perspective.

International norms

Before the Rancho Mirage summit, President Obama claimed that China had failed to abide by international norms in cyberspace. President Obama contended that '[t]he development of norms for state conduct in cyberspace does not require a reinvention of customary international law, nor does it render existing international norms obsolete' (International Strategy for Cyberspace 2011). Instead, he believed that '[l] ong-standing international norms guiding state behavior – in times of peace and conflict – also apply in cyberspace' (International Strategy for Cyberspace). In the meeting with Xi, President Obama – fully aware of China's status as a great power – asserted that China should play a role in shaping international norms and rules. However, the term 'international norm' in reference to cyberspace is quite vague: quite simply, what is an 'international norm' for cyber security? Koh (2013) suggests that, in developing international norms for cyberspace, the US should use 'diplomatic law talk' to relevant parties as a starting point to promote consensus among international partners, especially in the realm of cyber conflicts. Koh further suggests that

> the United States and other international community members should concentrate on fostering discussion and building consensus about a set of norms, rules, principles, and decision-making procedures that converge and apply in a particular issue area. Some of the documents that emerge from these diplomatic discussions might be described as 'soft law,' inasmuch as they seek to define new norms, or speak to how established norms should apply to new circumstances (Koh 2013: 742–3).

Indeed, in the publication of the International Strategy for Cyberspace, the White House stated that '[t]he United States will work with like-minded states to establish an environment of expectations, or norms of behavior, that ground foreign and defense policies and guide international partnership' (International Strategy for Cyberspace 2011). In the international context, a UN panel of cyber-security specialists also recommended that the international community should have further dialog on cyber-security norms (UN Secretary-General 2010).

From this perspective, the Obama administration might be unwise to accuse China of violating international cyber-security norms because China could easily rebut the accusation by pointing out that no such norms currently exist. Although the Obama administration has promoted the development of norms for respecting intellectual property and cyber-security due diligence, even President Obama himself admitted during the Rancho Mirage summit that international cyber-security issues are 'uncharted waters' (Sanger 2013). A better approach for the US government to take toward China is to get the latter to the table and build consensus on this issue. This is exactly what Obama aimed to do at the summit with Xi. Earlier in 2013, before the summit, the US National Security advisor proposed similarly that '[China should] engage with [the United States] in a constructive direct dialogue to establish acceptable norms of behavior in cyberspace' (Etzioni 2013). In December 2013, the United Kingdom similarly called for a dialogue with China over cyber-security issues (Watt 2013). As China and the United States are the most influential powers in the global political economy, any consensus that they reach would, in all likelihood, closely approximate international norms acceptable to a much broader group of countries. It is occasionally easier to build norms than treaties. Furthermore, consensus between the two powers would perhaps be much more effective than norms built by many other countries.

Bilateral agreement

The United States has initiated a series of bilateral meetings with a number of important powers, including China, to discuss the relationship between cyber security and critical infrastructure protection (CIP). However, the scope and the outcome of these meetings are obviously insufficient for the task of satisfactorily addressing US security concerns about China's aggressive cyber activities. At the Rancho Mirage summit, one achievement of the United States was to convince China to engage in regular bilateral discussions with the United States on cyber-security issues. Although there is no concrete agenda in the bilateral dialogue, it constitutes a major step forward for the two powers, which now have a forum in which they can exchange ideas on this topic.

Some commentators are quite pessimistic about the cooperation between China and the United States, because they view China as a country that does not cooperate much. For instance, although US Secretary of State John Kerry and Joint Chiefs of Staff Chairman Martin Dempsey traveled to Beijing to discuss this issue in early 2013, sources report that the People's Liberation Army has continued its aggressive cyber activities against the United States. As Tom Donilon states, both China and the United States were undermined by mistrust in the realm of cyber security. One particularly important task is to identify why the Chinese government would forego its engagement in any cyber-hostility activities or its possession of any cyber-hostility capabilities? Would the Chinese government be willing to exchange these activities and capabilities in exchange for a mutual limitation by the United States or other adversaries? As Goldsmith (2013) pointed out, embarking on a search for answers to these questions might lead to an unsolvable prisoner's dilemma.

It is probably incorrect to conclude that China is completely indifferent to international cooperation on cyber-security issues. In fact, China, Russia, Tajikistan, and Uzbekistan introduced the draft resolution entitled 'International Code of Conduct for Information Security' to the United Nations General Assembly in September 2011. It is surprising that neither Obama nor Xi mentioned China's 2011 proposal in the Rancho Mirage summit. Obama probably hesitated to refer to the 2011 proposal because it failed to substantively address security concerns of the United States, and it seemed to recognize the legitimacy of national governments' central control of online information flow (Whitt 2013; Farnsworth 2011). It is more difficult to conjecture why Xi did not mention the efforts undertaken by the previous Hu Jintao administration on international cyber security. Xi may have wanted to maintain a friendly atmosphere at the summit and avoid making a controversial statement given his knowledge of the intense criticism that the 2011 proposal had drawn from the US government.

Even though Xi and Obama ended the Rancho Mirage summit without acknowledging the 2011 International Code of Conduct for Information Security, this essay argues that the 2011 proposal may be an ideal starting point for the two main cyber powers to begin bilateral dialogue on cyber-security issues. Although the US government is disappointed with most of the content in the proposal, there remain some positive signals one can read from it. First, China agrees that international rules for cyber-security concerns are desirable, and it recognizes 'the need to prevent the potential use of information and communication technologies for purposes that are inconsistent with the objectives of maintaining international stability and security and may adversely affect the integrity of the infrastructure within the States, to the detriment of their security.'[2] Most important, the 2011 proposal can be read positively insofar as China, a political and cyber superpower, appears willing to facilitate international cooperation in setting new rules for cyber security. If the United States can recognize the preliminary efforts of China and their value, it will be much easier for the two nations to positively engage in bilateral discussions. On the other hand, it will be a great opportunity for the Xi administration to demonstrate that as a great power in the world, China is playing an indispensable and responsible role in setting the new global order.

One positive outcome of the Rancho Mirage summit is that the US-China governmental working group on cyber issues was first held in July 2013 after the summit (Gertz 2013). This inaugural meeting was framed as a part of the larger US-China Strategic and Economic Dialogue, where both the US and China reached consensus on increasing cooperation and building understanding on cyber-security issues. Although this seems to be a good start, China suspended its participation in the working group in May 2014 because of the US indictment of Chinese military officials on hacking charges (China Daily 2014).

Multilateral treaty

The increasingly rampant cyber conflicts and cyber attacks have created new challenges to current international law theory and practice, especially the concept of 'law of wars' (Graham 2010). Some commentators have proposed introducing an international treaty to counter international cyber-security threats. This proposal views cyber attacks as a type of military attack that undermines national security. Like all other international treaties, a cyber-security treaty would be successful only if the international community's principal actors would have the political will to craft a mutually satisfactory treaty and to enter into it in earnest. International cyber-security expert Kenneth Geers (2011) believes that any such convention should have the support of at least the United States, Russia, China, and the European Union.

The International Code of Conduct for Information Security proposed by China, Russia, Tajikistan, and Uzbekistan to the United Nations General Assembly in 2011 was a call for a multilateral UN-based framework leading to the establishment of international rules on cyber security. Although the proposal failed to garner the support of most UN members, the United Nations has become probably the most important international organization that can help build consensus on cyber-security issues in the international community. Cyber powers, including the

United States, China, and Russia, have engaged in relevant discussions at various points in time and have agreed that further dialogue is needed within the United Nations. As a permanent member of the United Nations Security Council and a widely recognized cyber power, China's role in cyber security will unquestionably have a profound effect on the future of related international norms and rules. If China and the United States can build consensus through a bilateral agreement, a multilateral treaty will be much more likely to take shape and succeed.

It stands to reason that cyber wars or cyber attacks are different from traditional wars or military conflicts. Occasionally hackers may route through 'cyber safe havens' or other countries. This is the so-called 'attribution problem.' For example, in 2009, a cyber attack brought down some American and South Korean websites. Initially, investigators concluded that North Korea was the geographical area from which the attack had been launched, but later evidence suggested that the attack had originated in Miami and had subsequently been routed through North Korea. In the end, it remained unclear who launched the attack and from where. Another example of the difficulties in assigning responsibility for cyber attacks concerns a case that took place in 2009, when the Information Warfare Monitor, a project supported by the Canada Centre for Global Security Studies in University of Toronto, revealed a plot called 'GhostNet' emanating from China and infiltrating more than 1,000 public and private computer systems in over 100 nations. Nonetheless, the report was unable to rule out the possibility that the masterminds of GhostNet were, in fact, hiding behind and routing through China. As IT expert Steve Armstrong suggested, it is possible that some other countries have conducted cyber attacks through China. Some cyber attacks, therefore, might be wrongfully attributed to China. All of this murkiness in assigning responsibility for cyber attacks raises significant doubts about whether the international community can effectively enforce cybersecurity treaties. If we optimistically conclude that enforcement is possible, then we must at least acknowledge the necessity of other supporting factors, including an enforcement system that features an elite professional staff, cutting-edge technology, and a robust international network.

Some international law scholars are not so optimistic, taking a dim view of international treaties on cyber security after considering the relevant political and technological circumstances. For example, Goldsmith (2011) argues that insufficient alignment of powerful nations' respective interests will prevent them from crafting an effective treaty on cyber security. This prediction may prove true in the near future. However, as nation-states and corporations increasingly gain awareness of cyber attacks' capacity for destruction, dialogue between different stakeholders in the international community will undoubtedly take place and shape future global Internet governance, whatever form it may take.

Conclusion

Having evolved from a technical discipline to a strategic and legal issue, cyber security is now playing an increasingly important role in international disputes – a trend that shows no signs of abating. The Obama administration's continual solicitation of China's collaboration in establishing and enforcing international cyber-security protocols not only reflects the importance of this issue in the international economy, but also proves that China has become both a physical and a digital power globally. Cyberspace undoubtedly provides an important dimension for observing China's current economic, foreign, and military strategies.

When the Nobel Laureate Liu Xiaobo (2006) praised the Internet as 'God's gift to China,' what he meant is that by fostering free expression and information flow, the Internet can help democratize the country. Ironically, Liu's statement takes on an entirely different meaning from the perspective of the Chinese government – the same government that he has been criticizing vigorously and that imprisoned him for years. For the Chinese government, digital technologies as applied to cyberspace are indeed God's blessings, helping China's leadership develop the national economy, control speech, and gain unprecedented power on the international stage.

As the US government is now pressing China to work collaboratively on setting new norms and rules for international cyber security, the Xi administration must decide on China's role in global cyberspace. Although we can hardly expect an ideal multilateral treaty on cyber security emerging in the near future, Xi Jinping and Li Keqiang have signaled their openness to cooperative, rather than confrontational, steps toward shaping international cyber-security norms. Ultimately, Xi's legacy in the history of the Internet will depend on how China, under his leadership, chooses between aggressive behavior and visionary collaboration in the global digital arena.

Notes

1 This paper has been previously published as 'The Red Storm in Uncharted Waters: China and International Cybersecurity' in *UMKC Law Review*, Vol. 82, No. 4, 2014.

2 International Code of Conduct for Information Security, U.N. Doc. A/66/359 (Sept. 14, 2011).

7
Cyber Operations in the Middle East

Jeffrey Bardin

Introduction

The Middle East cyber domain is as explosive as the current physical environment. Rife with constant change, new threat actors, virtual safe havens, and the expansion of extremist views are creeping into once protected virtual domains. The advent of social networking in the Middle East and elsewhere provides a real-time method for propaganda delivery and unstructured command and control. Social media is rapidly becoming the tool of choice for the dissemination of various types of information, propaganda, and calls to action. The ease of use and broad, rapid reach of messages make social media a logical choice for raising awareness and launching protests. Users of social networking tools such as Twitter and Facebook create accounts, which then get accepted or added by followers or friends, requiring specific foreknowledge or reputation-based access to the inner circles of the account holder. The utility of this communication method became apparent during the series of Arab Springs that stretched from North Africa through the Gulf states.

Multiple access vectors to the Internet combined with scores of ways to publish information led to a completely new strategy for information dissemination in the Arab world (Lotan et al. 2011). These methods of connecting and sharing information overwhelmed government propaganda centers and information control outlets. Gone were the filters that prevented free speech. Removed was the threat of discovery for openly communicating a dissenting view. This relatively open communication allowed the ready organization of protests, rallies, and political change. As open communication flourishes, so too does the opportunity for the spread of extremist opinions. These are opinions long since secured away in dingy dungeons, yet honed by years of refining and planning.

The advent of open communications protected by the relatively newfound anonymity and the speed of the Internet has created a novel circumstance in which the extremists, and others, now find themselves.

Social networking has become the Command, Control, Communications, Computers, Intelligence, Surveillance and Reconnaissance (C4ISR) for the oppressed (Defense Technology 2011). Social networking sites such as Twitter and Facebook serve as weapons in the communications and propaganda war. Communications through these media are the tip of the spear for mobilization, radicalization, and planning. All weapons are seen as threats. Governments throughout the Middle East initially attempted to prevent the use of these weapons but have shifted their approach. A new dimension in doctrine appeared in the form of social networking sites becoming targets for espionage and infiltration. Middle Eastern governments now infiltrate social networking sites with methods to disrupt, interfere with, and misdirect communications and planning activities. The overall rise of the cyber approach to both fomenting and combating extremism may be summarized as a direct reflection of the physical unrest in the Middle East driven by years of dictatorship-led repression and oppression (Bowman and Camp 2013).

Evolution of cyber strategies – Iran and Syria

Unlike the heavy financial burden of building armies, navies, and air forces, defensive and offensive cyber-warfare buildup may be established at a relatively low cost. In addition, there is no need to send troops to targeted geographic locations. The targets can be surgical in nature, electronically precise, and efficient. The results of targeting may be politically embarrassing and highly effective, leaving the target to wonder what comes next. Technical knowhow and an innate penchant for intellectual curiosity and exploration are starting points for building the foundation of a cyber-warfare strategy. The current evolutionary process for cyber-war capabilities follows an initially unstructured format that eventually matures into a structured and functional model backed by government and academic support. The steps in the process are described below:

1. Access to the Internet
2. Experimentation and exploration of vulnerabilities on adversary sites
3. Hacking as a 'script kiddie' with random defacements
4. Continued honing of hacking skills escalating to data exfiltration
5. Loosely formed organizations based upon nationalistic leanings

6 Participation in capture-the-flag activities
7 Limited creation of cyber-security companies
8 Academic institutions creating defensive and offensive curriculum
9 Locally held cyber-security conferences
10 Globally announced targeting of adversaries based upon religious and political views
11 Government formation of defensive cyber programs
12 Government formation of offensive cyber programs
13 Integration of loosely formed organizations with government entities and academic institutions
   a. Use of loosely formed organizations as proxies to target adversaries based upon government priority targeting requirements
   b. Outside virtual organizations mirroring physical proxies
14 Academic institutions expanding curriculum at the master's level
15 Academic institutions holding capture-the-flag games
16 Active cyber actions against adversaries during periods of political or religious unrest, or military engagement as an instrument of foreign policy
17 Government proclamations of cyber security as a national imperative

The experimentation and exploration of cyber vulnerabilities whets the appetite of stakeholders who want to protect their interest in cyberspace. Once acquired, these newfound skills must be tested, validated, and expanded into a verifiable craft that is repeatable and adaptable. A hacker's skills must be constantly adapted since the speed of change in social networking sites and adversary website security requires continuous learning and modification of tactics, techniques, and procedures (TTPs). Syria and Iran demonstrated the above process during the formation of the Syrian Electronic Army (SEA) and the Iranian Ashiyane Digital Security Group (Ashiyane 2014). Although not identical in methods, the two groups followed a path that eventually led them to integration with government and academic institutions. Out of necessity, nationalistic fervor and assumed forced march, SEA and Ashiyane aligned themselves directly with respected institutions.

Several other groups also demonstrated evolution in their sophistication and use of cyber capabilities, such as the Iran Babol-Hackers Security Team, Persian Crackers, Iran Hacker Association, and the Iran Hacking Sabotage Team (Smith 2006), as well as the Syrian Malware Team (Muncaster 2014). Another Iranian group, the Iran Cyber Army, is reportedly tied to past attacks on Western organizations (Bright 2011). Iran's Revolutionary Guard Corps (IRGC) has not openly claimed control over this group, yet multiple Iranian leaders have lauded their efforts as exceptional (Institut Francais d'Analyse Strategique 2012). The article goes on to name over 1,500 cyber-war commandoes in the Basij. The article also describes an Iranian cyber police force established to maintain the flavor, tone, and rhetoric of the 1979 revolution on Iranian-hosted web and social networking sites.

Iran has integrated cyber warfare into its military activities. Both the IRGC and the army have attempted to deal with US cyber signals and communications intelligence-collection capabilities. They do so by making extensive use of buried fiber optics and secure communications and developing more secure ways to use the Internet and commercial landlines (Cordesman 2007). Hossein and Sharif Universities are working on an 'impenetrable intranet communications network' (Cordesman 2007). Iran claims such a system was fielded during the Eghtedar exercises in February 2007 (American Intelligence 2007).

Integration of cyber warfare into educational programs and activities

Education is considered a top priority in Iran's plans for economic development. To this end, the government has worked to increase the primary education enrollment rate (Maslen 2013). Additionally, Iranian enrollment in US universities rose by 25 percent to 8,700, compared to a 25.65 percent increase for other Middle Eastern students (Maslen 2013). The increased emphasis on achieving higher education is producing the computer scientists and technology engineers necessary to have an advanced cyber weapons program (Carroll 2008).

Iranian academic institutions, such as Sharif University, are expanding academic program offerings into cyber operations, responding to a national imperative to protect critical infrastructure as well as develop methods to not only counter attacks but initiate cyber offensive actions of their own. Sharif University offers traditional information security and technology courses and has expanded into advanced topics such as network security (Advanced Network Security CE 40-817 - Kharrazi 2014).

A closer look at this topic reveals the need to establish a baseline for understanding security and privacy, with a standard taxonomy that provides insight into hackers and hacking (CE 817 Advanced Network Security 817-902 Lecture - Kharrazi 2012). In his courses, taught at Sharif Technology University, Professor Kharrazi teaches Iranian students how to recognize a system's vulnerabilities and threats, commenting: 'You cannot design a security system unless you know who the enemy is'

(CE 817 Advanced Network Security 817-902 Lecture - Kharrazi 2012). Regular lectures cover hackers, hacking organized and disorganized crime, industrial espionage, inside jobs, and spies, as well as traditional methods of attack. Much of the learning material is drawn from Western sources for proven academic material. Like most college courses of this type, lecture is coupled with hands-on labs. Professor Kharrazi covers lab activities such as Port Knocking, wireless assessments with WireShark, and Honeyd. Honeyd is an open-source honeypot. Students are tasked with creating two fake virtual hosts on both Windows and Linux using Nmap and Hping to scan the virtual hosts. The students are asked to track logging activities, comparing results with the honeypot configurations that are provided (CE 817: Advanced Network Security Homework 2 - Kharrazi 2014).

Capture-the-flag events

Iran's Sharif University currently sponsors annual capture-the-flag (CTF) events. The 2014 event included CTF areas covering web attacks, steganography, cryptography, forensics, secure coding, reverse engineering, and reconnaissance (Sharif University 2014). Translated into traditional tradecraft terminology, Sharif University is training students in offensive cyber operations, denial and deception, investigative techniques, cyber defensive measures, and cyber intelligence collection. The competition is international in flavor, but all write-ups (results of the CTF activities) are published for review and analysis. These analyses provide future educational material for all participating countries after the CTF activities are complete.

However, the size of the Islamic Revolutionary Guard Corps (IRGC) cyber warfare unit was reported to be 2,400 staff with a budget of $76 million (Lewis 2011). The tactics of the Shamoon attack on Saudi Arabia and Iran demonstrate increasing capabilities driven by foreign policy. Shamoon spread across PCs and shared network drives, wiping the drives and the master boot record of the devices to prevent boot up. The software reportedly derived its methods from previous malware left on the cyber battlefield without payload encryption.

Syria is not quite as advanced in its cyber capabilities but is learning directly from its allies, Russia and Iran. The Virtual Syrian University and Tishreen University represent examples of academic programs geared toward cyber operations and a workforce educated to protect Syrian critical infrastructure. The models in both Iran and in Syria have matured to the point where the loosely run groups of Ashiyane, Iran

Cyber Arm, and the Syrian Electronic Army no longer openly boast of their victories. This is presumed to be the result of new government oversight and control.

Several virtual organizations external to Iran that mirror physical entities are used as cyber proxies. Hamas and Hezbollah are two examples of Iranian proxies providing material support to Iran. Cyber organizations such as the Ezzedeen Al-Qassam Brigades target Israeli and Western interests, claiming to hit multiple Israeli sites in 2013 with the assistance of Anonymous, causing $55.4 million in damages (Ezzedeen AL-Qassam Brigades 2013). Syrian Electronic Army participation in this attack included e-mail instructions from its members documenting tool usage and targets (Treadstone 2014).

The Israeli cyber juggernaut

Israeli universities such as Tel Aviv University, the Israeli Cyber Defense Institute, and Ben-Gurion University have established cyber security and intelligence curriculum. Ben-Gurion University's curriculum focuses on traditional cyber security with courses in cryptography, network security, operation systems security, attack detection, security engineering, and security awareness development. The Israeli Cyber Defense Institute is teaching courses on Linux, security architecture, engineering security, Windows hardening, development, governance, management, and security projects. There is also a course in hacking techniques. Tel Aviv University is touting these programs, but there is little open evidence of the syllabus and curriculum. In fact, definitive descriptions of cyber operations, cyber intelligence, and/or offensive cyber curriculum is nowhere to be found. It is not unusual to obfuscate this information as a method of protecting the course material and subsequent capabilities of newly minted graduates.

Israel has long been on the cutting edge of cyber security, establishing a thriving startup community. Many of the startups find their roots in the Israel Defense Forces and organizations such as the Israeli Intelligence Unit 8200 (eight two hundred) that focus upon cyber intelligence, espionage, and sabotage operations. Israel is one of the world's leading cyber-security locations, on par with Silicon Valley and other US areas of research, development, and startup activities. Strong investment from the US has created centers of excellence, including the city of Beer Sheva. Companies such as EMC, IBM, and Lockheed Martin have invested heavily in Israeli venture capital, establishing Beer Sheva as the cyber-security hub of Israel. Prime Minister Benjamin Netanyahu has driven investment and invested much energy in marketing this area through the National Cyber Bureau, the launching of CyberSpark, and the Israeli Cyber Innovation Arena.

The National Cyber Bureau serves to advise the prime minister on three central areas (National Cyber Bureau 2011):

1 Advancing defense and building national strength in the cyber field
2 Building up Israel's lead in the cyber field

Advancing processes that support the first two tasks, a closer look at Israeli government's establishment of the Cyber Bureau through Resolution No. 3611 demonstrates a commitment to end-to-end support:

3 'To advance coordination and cooperation between governmental bodies, defense community, academia, industrial bodies, businesses, and other bodies relevant to the cyber field' (National Cyber Bureau 2011).

The quotation above is clear as to the requirements needed to create a complete and holistic national approach to cyber security. As with most all governmental documents of this type, there is no direct mention of offensive cyber operations. However, Israeli intelligence has included cyber in their specialized unit capacities. There are some indications that Unit 8200 is focused upon signals intelligence; Unit 504 focuses on human intelligence and interrogation; Unit 5114 conducts radio surveillance; the Hatzvar Unit is focused on open-source intelligence and language (Global Security 2011); Bahad 15 is focused on intelligence tradecraft; and the IDF Information Security Unit targets computer systems and confidential documents' security.

Real-world cyber operations

There has been a steady stream of cyber attacks executed within the Middle East over the past few years. Below is a listing and description of attacks involving the three countries covered in this chapter based upon data collected from Hackmageddon (www.hackmageddon.com).

The Syrian Electronic Army hacked the main Harvard University website on 26 September 2011, defacing the site with a message of having been there. On 17 January 2012, an attack attributed to the IDF Team takes down the Saudi and UAE stock exchange websites. It is assumed that IDF stands for Israeli Defense Forces (or some entity representing itself as such), but this is not confirmed. On 20 January of the same year, TheJ0k3rS, an Israeli hacking group, reports to have taken control of several Iranian websites and releases a video demonstrating the hack. In a defensive move for training, the Israeli National Cyber

Command held its first cyber terror drill called 'Lights Out,' which simulated attacks on critical infrastructure, on 25 January 2012. As a possible extension of this exercise, on 26 January, the IDF took down several Iranian government and media sites, leaving a post of the Israeli flag in their place.

In response to the focus on cyber security by both Tel Aviv University and the National Cyber Defense Authority, unknown Middle Eastern hackers penetrate the site and leave images of burning US and Israeli flags. Although still in the site-defacement genre, the political embarrassment is significant for the Israelis. In early 2012, the government of Iran announced several messages and initiatives in the cyber realm. The pronouncement that its nuclear facilities are immune from cyber attacks in light of past attacks from Stuxnet and Duqu was followed by Iran's supreme leader Ayatollah Ali Khamenei announcing the formation of a supreme cyberspace council (Iran Media Program 2013). Iran then held its first national conference on cyber defense in which Behrooz Kamalian of the Ashiyane Digital Security Group was a featured speaker. Flexing its newfound cyber muscles, the Iran Cyber Army attacks the Israeli ally Azerbaijan, leaving messages about being the servants of Jews. This attack was followed by the Iranian proxy the Hezbollah Cyber Army attacking two Iranian reformist and anti-government websites. The most significant attack of 2012 in relation to the target countries was the Shamoon attack of 15 August. Saudi Aramco was hit by a self-replicating virus that spread across more than 30,000 Windows devices, wiping the hard drives and master boot records of the devices and rendering them useless (Acohido 2013). The sophistication of this malware placed Iran on the map as a force in cyber warfare.

Progressing into 2013, Iran began to demonstrate its cyber espionage and sabotage prowess. A series of attacks against major US banks in the form of a distributed denial of service attacks used data centers from around the world infected with malware named 'itsoknoproblembro' architected to evade antivirus protection (Peterson 2013). Iran reportedly hacked non-classified US Navy computers in an attempt at cyber espionage in September of 2013.

The Syrians, via SEA, were able to pull off an impactful hack of the Associated Press's Twitter account, publishing a false news report about an explosion at the White House that caused a short-term decline in US stock prices. On 2 September 2013, we saw the SEA hack a US Marines website, again in a defacement exercise reminiscent of script kiddie activities. In 2014, the SEA hacked the ISD official blog. This was followed by a hack of an Israeli Army Twitter account that posted a warning message about a nuclear leak in Dimona, an Israeli city that's home to a nuclear research facility (Chasmar 2014).

Growing maturity

The series of hacks described above demonstrates a maturation of hacking capabilities and the use of targeted payloads. It is obvious by the types and methods of attacks that Iran has progressed far beyond the capabilities of the SEA, adapting the same holistic approach as the Israelis. Though the Syrians have been preoccupied with internal strife, they have been able to perform hacks of various targets. For Syria, the intent during this early stage of cyber warfare is as follows:

- To publicly 'strike a blow' against a perceived enemy
- To embarrass a targeted site by illustrating a security issue
- To attract public attention to a cause, an 'injustice,' or an entity
- To challenge/deny informal web server use by an organization
- To reduce public confidence in the security of a system and its trustworthiness for use for sensitive purposes
- To force a targeted system to be taken out of service until it can be scrutinized/analyzed, formatted, rebuilt, and hardened
- To establish 'street cred' with one's hacker/cracker peers, or simply because the defacer finds doing defacements to be 'fun'

Regardless of the intent, the skills and experience gained during these events serve to broaden and embolden the groups involved.

Cyber weaponry

There are many existing types of cyber weaponry in use. The world has seen the likes of Stuxnet and Duqu as well as Flame and Shamoon. Weaponry left on the cyber battlefield has led to innovative modifications and enhancements. Weaponry with unencrypted payloads leads to payload reverse engineering just as physical weaponry captured or acquired during combat operations leads to adversary discovery of unique technologies. Similarly, the distribution and movement of small arms around the world is mirrored in the spread of cyber weapons. In a 2003 report (Small Arms Survey 2003), at least 1,134 companies in 98 countries worldwide were involved in some aspect of the production of small arms and/or ammunition. In addition, massive exports of small arms by the US, the former Soviet Union, China, Germany, Belgium, and Brazil during the Cold War took place commercially and to support ideological movements. These small arms have survived many conflicts and many are now in the hands of arms dealers or smaller governments who move them between conflict areas as needed. Past activities of Anonymous demonstrate the same type of small-arms proliferation, albeit on a virtual plane.

Driven largely by ideological activities, the Internet collective Anonymous distributes a revamped version of the Low Orbit Ion Cannon (LOIC) tool used in mass distributed denial of service (DDoS) attacks. LOIC was the primary weapon used by Anonymous in its ongoing Operation Payback DDoS campaign against film and recording industry associations, as well as other organizations involved in anti-piracy efforts (Mansfield-Devine 2011). The application was originally created by a user named Praetox and was used in several mass attacks over the years, including Anonymous' campaigns against the Church of Scientology, Australian government websites, and the Iranian election protests in 2009. In January 2009, the code of the Windows program was released on Source Forge, an open-source community, as an open-source project. This enabled a cross-platform Java version to be subsequently created. A release such as this allows for the proliferation of code that can be enhanced, improved, and utilized in low-intensity conflicts with the potential for significant media coverage.

In 2013, another developer branched off that same code and added a new feature called 'Hive Mind' to the tool. This feature allows users to relinquish control over the application after installation and makes it act as a botnet client, which can be controlled from an IRC channel. This method of virtual arms proliferation allows like-minded individuals to participate in DDoS activities based upon their ideology while giving up control to centralized resources. Infiltration of Anonymous by Middle Eastern nation-states suggests an ability to influence and direct attacks against adversaries, serving as an unwitting proxy used as an instrument of foreign policy.

Small arms and light weapons have been responsible for the majority of the combat deaths in recent wars and figure in much of the crime and civil violence visited upon vulnerable societies around the world (Boutwell and Klare 2000). Virtual small arms are currently responsible for all the malware activities around the world today. This will continue for the foreseeable future and will be a constant source of frustration for most governments and many private citizens. The proliferation of virtual arms allows specific groups to more easily exact virtual revenge, practice censorship, and disrupt the activities of those who disagree or have a different ideology. The virtual small-arms trade also provides training opportunities for countries such as Syria and Iran.

Virtual small arms are ideal methods for online disruption. They are widely available and are low in cost or even free. They deliver a strong payload that is simple to use, highly portable, easily concealed, and potentially possesses legitimate military, police, and civilian uses (Boutwell and Klare 2000). These virtual weapons are light in footprint and thus can be used by the very young and technically astute that have played such a significant role in recent virtual conflicts.

Once the virtual conflict is over, virtual small arms still exist in the hands of the participants. Virtual small arms can easily be used to start other conflicts that may be more personal. This creates a surplus of virtual small arms, establishing a culture of hacktivism and an endless circle of virtual conflicts. Ashiyane, SEA, and the Iran Cyber Army have benefited from the virtual conflicts of others. Through the infiltration of organizations such as Anonymous, these organizations have enhanced their education in tactics, techniques, and protocols. These groups also benefit by acquiring virtual small arms from the cyber battlefield.

A past concern was that if Anonymous acquired much of the code related to Stuxnet, this ensured nation-states could acquire the code as well. According to some experts, malware is largely uncharted territory for Anonymous, 'which has built its notoriety on crippling the websites of governments and multinational corporations, such as Visa and MasterCard, which it deems a threat to freedom of speech' (Halliday 2011). The problem is that no one really knows the capabilities of Anonymous, since it is such a loose-knit group that comes together to crowdsource their targets and attacks based upon a shared ideology or belief. Anonymous uses Web 2.0 technologies to establish a community-based design for their focused efforts. They also use it as a method to propel their payload to new levels. The methods in use by Anonymous represent a virtual classroom for Syrian and Iranian government and non-government organizations. The residual virtual small arms left over by Stuxnet and other virtual small arms provide a foundation for melding together new attacks that can be leveraged in much the same way that LOIC has been leveraged and matured over time. The asymmetrical methods of cyber hacktivism used by Anonymous provide a cyber warfare firing range serving as a testing and proving ground for new attacks.

A list of existing cyber weaponry compiled by the Pentagon provides insight into the growing sophistication and lethality of attacks (Nakashima 2011). Spam and phishing may serve to provide initial infiltration of targets. Once exploited, new payloads specifically crafted based upon intelligence gathered is downloaded to the target site. The payloads may enhance cyber-intelligence techniques, include methods of intelligence and sabotage, or may be specific to certain operating systems and hardware. The payloads may serve as the initial data-collection mechanism, embedding sleeper software to be activated in the event of physical or cyber hostilities. Payloads may be lethal from a kinetic perspective, generating a physical outcome due to cyber activity. Payloads may also have a non-lethal, non-kinetic effect, denying an adversary access to their cyber weaponry.

Cyber weapons have attacked the domain name system, border gateway protocols, and fundamental Internet protocols. Botnets have been used to execute distributed denial of service attacks. SCADA systems, as well as other critical infrastructures, have been targeted. An examination of CTF activities in Iran shows each of these weapons in use during the CTF games.

United States positioning and preparations

The United States is well positioned as one of the foremost leaders in cyber-warfare capabilities. The FY 2013 Congressional Budget Justification for National Intelligence listed five guiding principles woven throughout the budget (National Intelligence Program 2012):

1 Sustain a skilled workforce.
2 Focus on further integration and collaboration.
3 Bolster agile capabilities that support multiple missions.
4 Enhance counterintelligence.
5 Protect key investments for the future.

Although there are drawdowns and cutbacks defined within the budget, key areas of continued growth represent areas within the cyber-intelligence lifecycle. Signals intelligence, cyber security, counterintelligence, and mission-focused science and technology all have cyber components. The budget request demonstrates a focused commitment to cyber-warfare capabilities.

Presidential Policy Directive 20 (PPD20) defines the US government's direction in developing, maintaining, and growing cyber-warfare capabilities. PPD20 clearly defines 'offensive cyber effects operations,' 'cyber operations with significant consequences,' and 'response to persistent malicious cyber activity' as growth areas (The White House 2013). 'Offensive cyber effects operations' capabilities focus on cyber operations against adversaries that may cause subtle to severe damage (The White House 2013). 'Cyber operations with significant consequences' require presidential approval, while 'responses to persistent malicious cyber activity' refers to mitigation and retributive counterstriking based upon insufficiency of network defense or law enforcement (The White House 2013). Both PPD20 and a budget justification for National Intelligence call for improved data sharing across federal and partner organizations, a critical aspect of a successful and integrated approach to cyber warfare.

The United States must continue developing a strong, educated cyber workforce by integration with the academic community. Many programs for cyber security, cyber operations, and cyber intelligence are in place at both the undergraduate and graduate levels. Commercial training programs exist to train cyber-war fighters and open-source collection specialists. Cyber intelligence courses are available to government, military, and commercial entities. The courses and training programs may or may not openly claim offensive cyber techniques as part of the curriculum. However, many of these programs do teach methods of denial and deception, degradation and disruption, penetration and manipulation as well as the ability to destroy computer systems and information. The public-private approach to cyber capabilities ensures a trained workforce. CTF activities and cyber-warfare ranges ensure weapons are thoroughly tested as part of a comprehensive training program and serve to provide a 'spare parts inventory' of targeted software and hardware. All cyber-warfare capabilities undergo a weaponization process before the capability can be validated as a cyber-weapons system. The US funding of cyber-warfare ranges demonstrates a commitment to cyber weaponization. Compared to physical weaponization, cyber weaponry may have a very short shelf life due to the rapidly changing cyber environment. This requires continued development of weaponry capabilities in an iterative method, as opposed to past physical weapon lifecycles that may take years to develop and deploy. The cyber-war fighter may also be seen as part of the weaponization process, since keystrokes performed by the war fighter alone may serve to infiltrate, sabotage, and damage a target system. The cyber-war fighter may only be armed with training, education, and hands-on skills, but these can still be used in a weaponized manner. The US cyber warfare weapon system should have the capability for command and control, testing, defined safeguards, delivery methods, trained personnel, tactics, techniques and protocols, and various attack vectors and launch platforms (US STRATCOM 2009).

Conclusion

The ubiquity of the Internet, availability of cyber weapons capabilities, and concerted efforts to train and educate the workforce all contribute to the current cyber arms race. Whether in Syria, Iran, or Israel, long-range planning with clear goals and objectives is needed to develop and deploy a fully integrated cyber-warfare capability.

US adversaries are rapidly establishing the infrastructure needed to deploy their own information operations capabilities and have demonstrated their prowess in doing so. The US must create a total force, cyber-warfare program ensuring integration with physical functions that is available on demand and globally accessible. The US must continue to develop a well-trained workforce that has authorized access to the tools and facilities necessary to carry out foreign policy objectives. The ability to gather, organize, and produce data, while analyzing it into socially and culturally integrated actionable intelligence, should be considered a key factor in the realm of cyber warfare. It is the intelligence function that lays the foundation for future actions.

Middle Eastern adversaries are numerous and are rapidly learning to use US-generated intellectual property and methods. From script kiddie experimentation and hacktivist activity to CTF and government-sponsored programs, cyber operations functions are maturing at a rapid rate. The US programs need to be agile and built on the premise of rapid change and adaptability in order to sustain the pace and sustain its technological edge. Cyber capability development programs need to be part of a national effort, with federally aligned education feeder programs that begin in grade school and continue through university studies. Social, cultural, linguistic, religious, and technical curriculum is required to effectively target our Middle Eastern adversaries. This type of curriculum enables an understanding of the adversary that is second hand and not a subject-matter-expert-only realm of knowledge. Information sharing should expand beyond just federal entities to private organizations with the skill and experience to follow foreign policy objectives within given bounds and rules of engagement. The US is well positioned to execute upon these imperatives. Doing so ensures the ability to effectively react to existing and emerging cyber threats from the Middle East.

Section III

Defense Strategies and Practices

8

A National Strategy for the United States Cyberspace

Harold 'Punch' Moulton, James Stavridis, and Constance Uthoff

Introduction

Over the past decades, the United States has become increasingly dependent on cyberspace in ways that offer both promise and substantial risk. The lives of Americans have improved, and will continue to improve, at an astonishing rate through the innovation, ingenuity, and broad deployment of cyber technology. On the other hand, there are a myriad of challenges. Threats lurk from the kid who just developed a new piece of malware that vandalizes your smartphone to criminals stealing high volumes of credit-card information to the highly sophisticated nation-state actions that put our society and national security at risk. Furthermore, how we cope with difficult challenges such as privacy demand a new approach to our collective future. The stark reality is that we have been on a ride where technology and innovation have been pushing the gas pedal to the floor, but we don't have anyone at the steering wheel.

During the last century, the United States have developed an FAA, NASA, and an air force for a new world with aviation, and we argue that during the 21st century the US needs a national organization to comprehensively advocate, organize, facilitate, and protect its capabilities in cyberspace. The discussion concerning government leadership in cyberspace is most often a debate over whether we should have the Department of Defense or the Department of Homeland Security in charge. The simple answer is neither. Both have missions far broader than just cyberspace, and despite their good intentions, neither department (nor relying on an interagency approach) would be able to accomplish what's needed for cyber moving forward in this century. We contend that the US needs a third option, a new organization singularly focused on cyber across all its many facets – the technical, the risks, the challenges, and the opportunities.

Merging the equities of numerous stakeholders in cyber is not easy. The domain is dominated by the private sector. However, the commercial and private world does not have the charter (nor the skill) to look out for the breadth of our national interests. That charter belongs to the federal government. Although a question subsists: Can we find a means to blend the strengths of the private sector with the charter of the federal government? In this chapter we argue that the answer should be 'yes.' A new quasi-governmental organization is the answer: one with a commission to ensure free and open access to cyberspace (the connected digital world), enable the conditions for growth in and through cyberspace, create a shared threat awareness, and form a common security partnership. Modeled on concepts from our Federal Reserve, a new Federal Cyber Board (FCB) can leverage private and public expertise with a national mandate to enhance our nation's competitiveness and security in the cyber arena.

The need for a national strategy

In the United States, dependence on cyberspace has been increasing for decades. Cyberspace, defined as 'the interdependent network of information technology infrastructures that includes the Internet, telecommunications networks, computer systems and embedded processors and controllers in critical industries' (The White House 2009), offers vast potential for economic growth and improvement in our quality of life. It 'includes broadband networks, wireless signals around us, the local networks in our schools and hospitals and businesses, and the massive grids that power our nation. It's the classified military and intelligence networks that keep us safe' (NSPD-54, 2008). It is much more than that. We are also interconnected by the 'Internet of Things' (IoT), where connectivity extends to devices such as refrigerators, insulin pumps, heart implants, smart homes, cars, offices, grids, and even e-cigarettes are all designed to offer simplicity and efficiency to daily life. In the end, cyberspace is 'ones and zeroes,' and they are nearly everywhere.

Today's digital age mirrors other great milestones of technological improvement that significantly changed the way we live, think, and interact. The printing press, telephone, aircraft, and television have transformed our societies. These technologies changed our communication, our perceptions, and ultimately the way we interact as a society. Similarly, cyberspace has introduced a new frontier. It's influencing how we see the world, how we relate with each other, how we relate with devices, and ultimately it is redefining how we live. Mobility, the cloud, and the IoT are vastly revolutionizing our world and will be expanding rapidly into the future. According to a report released by Microsoft (2014), we should expect 80 percent of Internet connections to originate from a mobile device by the year 2025. At that time, 'most of the data created in the world will move through or be stored in the cloud at some point and there will be upwards of 4.7 billion people online of which 75 percent will come from emerging economies' (Microsoft 2014). Moreover, Cisco anticipates that as early as 2020, there will be over 50 million connected devices (Evans 2011).

In the near future, the potential expansion of connected and cyber-physical systems (CPS) will increase the exchange of ideas, benefits related to health care, logistics, decision-making, capacity planning, and overall quality of life. A Foundation for Innovation report reflects that there is a growing trend toward computational intelligence, automation, and control for complicated but well-defined tasks or processes, especially when demands or constraints are not amenable to human intervention. For example, automatic collision systems could detect moving objects and respond faster than a human operator. Unmanned CPS could be used to reduce the risk to human life by detecting mines, exploring volcanoes, or conducting otherwise hazardous tasks. In the future, CPS could make possible concepts only imagined today, such as unmanned tours to the moon, bionic suits, and automated large-scale indoor agriculture system (Sztipanovits and Ying 2013). Advancement in technology will help to bring new products to market, driving opportunities for employment and economic growth. It is estimated that by the end of this decade, computer components will be part of more than half the value share in various sectors including the industrial and medical sectors (Sztipanovits and Ying 2013).

In terms of market growth, cyberspace will drive every corner of our economy. According to a 2012 report from General Electric, by 2025 the technical innovations of the Industrial Internet could value approximately one half of the global economy (Evans and Annunziata 2012). From things to services, cyberspace will be the fuel that makes our economy run. But is our society ready? How are we shaping a workforce to not only operate in this new world but lead it to new heights? The reality is that we are woefully behind. Education and training are not keeping pace. We deserve better, and the next generation deserves better.

Cyberspace is certainly poised to offer a broad range of opportunities, but, unfortunately, there is a dark side: from the Sony attacks to the

Target breach, cyber threats ranging from the annoying to the dangerously sophisticated have the very real potential to debilitate our way of life due to inherent vulnerabilities. One unique feature of cyberspace is its ability to provide anonymity, opening the gate to criminal and highly disruptive activities. From vandals to criminals to nation-level aggression, malicious activity abounds. Lone hackers, criminal organizations, collectives like Anonymous, proxies, and nation-states threaten our bank accounts, our intellectual property, our privacy, and our critical services. President Obama, during the 2013 State of the Union address, acknowledged the problem:

> We know hackers steal people's identities and infiltrate private email. We know foreign countries and companies swipe our corporate secrets. Now our enemies are also seeking the ability to sabotage our power grid, our financial institutions and our air traffic control systems. We cannot look back years from now and wonder why we did nothing in the face of real threats to our security and economy (The White House 2013).

More ominously, in 2014, Admiral Mike Rogers, director of the National Security Agency, declared he 'fully expect(s) that during (his) tenure as commander of the US Cyber Command there will be offensive activity directed against critical infrastructure of the United States designed to damage, destroy, or manipulate' (Reuters 2014). While we see growing awareness of the magnitude of the cyber threat, not enough has been done to collectively address it. Must it really take 'an electronic Pearl Harbor' to prompt significant change?

However, risks are not only cyber-security related, but are also tied to a host of economic and social challenges. The rate of technological change and the expansion of global connectivity of individuals and marketplaces will offer other challenges that the United States is not yet prepared to meet. According to the World Economic Forum (2014),

> Risks are not just from the commonly recognized sources – such as criminals, malware, or even state-sponsored cyber-attacks; they can emerge from policies as well. Societal responses to immigration challenges, education and workforce needs, trade liberalization, as well as international cooperation to resolve cyber conflict, will shape the future of cyberspace for both developed and emerging economies.

Cyberspace is testing some of our very core beliefs. For example, what does privacy mean in a world where facial recognition can be tied to security cameras at a store or commercial building? Is your conversation private if you text it over a cellular network? How about through Facebook? These challenges need some rigorous thought at the national level. Unfortunately, change may not happen quickly enough to address these and other emerging issues in cyberspace, a point illustrated by a report from the Center for Strategic and International Studies (CSIS) Commission on Cyber Security for the 44th Presidency. It argues that the authorities that are critical for effective cyber security are 'increasingly outdated' as a result of the fast rate of change in cyberspace (Lewis 2008). Similarly, according to a report from Congressional Research Services (2013), consensus has been growing that the current legislative framework for cyber security might need to be revised in order to address the demand for improved cyber security, especially considering how quickly technology and the threat landscape are expanding and evolving.

In short, we have two fundamental cyber concerns as a nation: one connected to empowering the opportunity of cyberspace, and the other addressing our risks from cyberspace. The direction and the outcome of how we blend these two vital vectors will shape our success as a nation in this century. For this to be possible, the United States needs a focused strategy that encourages private-sector growth and innovation while supporting flexible but efficient security and risk-management solutions. Currently, the United States lacks one cohesive national cyber strategy, organization, or agency to lead in this space. In order to capitalize on opportunity while minimizing risk to national security, it will be imperative to have a strategy or agency driving the initiative. A brief examination of strategy, authorities, and roles will help to illuminate current gaps and illustrate the need for change.

In 2013, the Government Accountability Office (GAO) released a study titled 'National Strategy, Roles and Responsibilities Need to Be Better Defined and More Effectively Implemented.' As a result of an increase of cyber-security incidents in the public and private space that could have potentially devastating consequences, the GAO examined the federal approach to cyber security, in part by reviewing national cyber strategies and their implementation. This report acknowledged that there is not one overarching US cyber strategy, and the ones that exist generally lack metrics to measure effectiveness.

> Although the federal strategy to address cyber security issues has been described in a number of documents, no integrated, overarching strategy has been developed that synthesizes these documents to provide a comprehensive description of the current strategy,

> including priority actions, responsibilities for performing them, and timeframes for their completion. Existing strategy documents have not always addressed key elements of the desirable characteristics of a strategic approach (GAO 2013).

The national cyber strategies, though necessary for improving federal cyber-security initiatives, are not focused enough or sufficient to uniformly prepare the United States for future challenges.

Luiijf, Besseling, and De Graff (2013) explore the concept of national strategy in their paper 'Ten National Cyber Security Strategies: A Comparison.' They define national cyber-security strategy (NCSS) as a national plan of action based on a national vision to achieve a set of objectives that contribute to the security of cyberspace. To these authors, national strategy aligns the 'whole of government,' unifies and clearly outlines roles and responsibilities among all stakeholders, and communicates a nation's cyber position and intentions. They emphasize that it is important to coordinate the full range of stakeholders across the national space, including the government, military, academia, citizens and regulatory bodies, and the public and private sectors.

Although Luiijf, Besseling, and De Graff (2013) argue that the highest priority that should come from a NCSS is international collaboration, they also note that a NCSS should relate to other national strategies, and they emphasize that an NCSS should be clear and align stakeholders with a common vision in a common direction. They further compare various national cyber strategies in areas such as vision, objectives, guiding principles, and action items. They also offer recommendations on how to create a NCSS based on their findings.

The paper also offers some interesting insights to the United States Cyber Strategy of 2003. The authors note that the National Strategy to Secure Cyberspace does not mention public–private partnerships or information sharing as key objectives. It also lacks 'specific, measurable, achievable, realistic and timely (SMART) definition of the actions' (2011). The strategy is meant to be a framework to help Americans to secure their own areas of cyberspace; however, it is not the only approach that has contributed to US national direction and vision. Since 2000, there have been at least 12 related national strategies as well as other related agency strategies; over time, there has been a collection of documents that have contributed to national strategy, emerging as circumstances dictated. Some include the 2011 International Strategy for Cyberspace, The Department of Defense Strategy for Operating in Cyberspace, The National Strategy to Secure Cyberspace released by the Department of Homeland Security, and National Strategy for Trusted Identities in Cyberspace. In 2008, through the National Security Presidential Directive 54 and Homeland Security Presidential Directive 23, the executive administration under George W. Bush established the Comprehensive National Cyber Security Initiative (CNCI). In 2013, along with 13636, the Obama administration released Presidential Policy Directive 21.

Perhaps to be expected, with so many strategies, orders, and directives, it has been difficult to create a cohesive national vision related to cyber innovation and security that includes collaboration from all stakeholders. Some of these challenges were discussed in the GAO report. Gaps included a decrease in compliance related to risk management required under FISMA in both the federal and critical infrastructure; DHS warning and timely analysis is not fully mature and is not expected to be until 2018; there are large gaps in federal workforce development, and the government still has limitations related to defining and implementing international approaches to cyber security (GAO 2013).

Interestingly, much like the Luiijf, Besseling, and De Graff (2013) paper outlined, the GAO report illustrates that US strategies lack performance measures, milestones (SMART actions), clearly defined roles and responsibilities among stakeholders, and alignment with other national strategies. According to the GAO (2013), as US strategies emerged, some failed to connect to previous strategies or to illustrate how they may fit into an overarching national cyber strategy. Though the office recommended that the White House cyber-security coordinator should create a more effective strategy that addresses the various gaps in authorities and ambiguities related to roles, responsibilities, and accountability, the Executive Office of the President, while agreeing that progress was needed, was not in favor of creating a new strategy. Unfortunately, without a unifying strategy or authority, it will be difficult, if not impossible, to effectively manage the risks and opportunities related to the future in cyberspace.

The US federal government seems to be uniquely skilled and positioned to address these types of cyberspace issues at a national level. However, much like the national cyber strategies, authorities are not clearly focused to meet future cyber concerns and growth. As noted by the GAO 2013 report, it is tremendously difficult to coordinate cybersecurity activities across the federal government in the absence of clear authority 'unless federal agencies institutionalize a coordination mechanism that engages all key federal entities, it is less likely that federal agencies will be aware of each other's efforts, or that their efforts taken together, will support US national initiatives' (GAO 2013).

In regards to agency-specific authorities, there is not one agency poised to tackle the issues related to opportunity and security cohesively. Currently, the Department of Homeland Security is the lead agency and has the primary authority for civil-sector cyber security. The National Security Agency (NSA) and US Cyber Command are the lead agencies responsible for national security and military cyberspace operations, respectively. The Department of Justice is largely responsible for the enforcement of laws relating to cyber security. In 2009, the White House cyber-security coordinator position was created to provide oversight of the government-wide cyber-security coordination. Other federal agencies function as diffused security and equity partners with cyber-security responsibilities involving their own systems, with additional cyber responsibility to their related critical services sector. For example, the Department of Commerce is involved with the National Institute of Standards and Technology (NIST) and the National Telecommunications and Information Administration (NTIA) in collaboration on the National Strategy for Trusted Identities in Cyberspace (NSTIC). Cyber authorities spread out among agencies and trying to measure the effectiveness of the federal government approach to cyber-security agency roles can be highly challenging to sort through. Additionally, gaps in legislation have created a vacuum that some agencies, despite the lack of official authority, are filling. At times, agency efforts are duplicated and, as a result, cross-agency coordination and clarity of roles can be complex.

Not only are roles diffused, but also there is not one agency that seems to have the capability to address future challenges either as a result of legal constraints, workforce challenges, or resources. For example, though the Department of Homeland Security is tasked as the primary national cyber-security agency for protecting the infrastructure, some question if it should remain the lead. It certainly has the authority through presidential directive, but without some additional developments related to workforce and authority, it has been difficult to measure its effectiveness. Fleming and Goldstein recognize in 'An Analysis of the Primary Authorities Governing and Supporting the Efforts of the Department of Homeland Security to Secure the Cyberspace of the United States' that though DHS is charged with the authority to lead in this space, they may not have the full capability to do so; they are limited in their ability to facilitate the protection of critical systems, and they lack clear lines of authority related to cyber attacks and the responsibilities of the Department of Defense (2011).

Jeff Moss, the founder of DefCon, but also a member of the DHS Homeland Security Advisory Council, reinforces this perspective. Moss argued that the Department of Homeland Security is the suitable entity for most cyber-security activities, but it does not have the authority to overhaul the United States' effort in cyber security, translating into a lack of human resources (Krasny 2014). Not only is DHS not staffed for it, but also there is significant turnover in DHS, which further impact its ability to keep ahead of emerging threats.

Another possibility is that the Department of Defense (DoD) could be the agency to take the lead. They have the capacity to defend their own systems, and they rely on many contractors for support. The DoD has the talent, the experience, and the support of NSA. The DoD clearly has the ability to protect cyberspace; however, it does not have the authority. Very simply, the Posse Comitatus Act limits the legal authority for the DoD to act within the United States in that capacity.

Without agency authority, cyber legislation could help to shape future roles and responsibilities. Unfortunately, legislation is also lagging. In the United States, attempts to pass cyber legislation have been mostly unsuccessful in recent years and, to date, there is not one specific legislative framework to guide national cyber-security efforts. In fact, no comprehensive cyber-security legislation has been passed since 2002. Without clear legislation, other agencies may seek to fill the cyber gap without authorization. In that respect, over recent years, without legislative authority, the FTC has taken a lead in cyber-security enforcement. Challenges to this were illustrated in the case against Wyndham Hotels regarding the breach perpetrated by Russian hackers leading to a $10 million fraud.

In response to FTC regulation regarding a cyber breach at a Wyndham location, Wyndham lawyers argued that cyber-security regulation should be developed through the legislative process, not by bureaucrats. Congress, they cited, not the FTC, has the authority over data-security standards. They maintained that current federal statutes authorize other agencies to regulate cyber security, not the FTC. Some fear any increase in FTC activity that tries to enforce cyber-security standards could be damaging, not only to industry but also to the overall government-led effort to coordinate cyber-security information sharing. Furthermore, private-sector firms have been pushing back, and though the FTC is gaining momentum as a federal lead in cyber security, it lacks legislative authority (O'Toole 2014). This has left wide gaps and ambiguity in relation to uniform standards and enforcement. Private-sector inclusion is important, and this could further strain the public–private cooperation that is imperative to meet national cyber-security threats.

Currently, there is no unifying strategy or agency that is in charge. Though the United States has a handful of much-needed national cyber strategies, they have proven to be somewhat disjointed and seem to lack the incentives for necessary collaboration and cooperation between the private and public sectors. To meet the future challenges and opportunities in cyberspace, in an environment that includes layers that are non-linear, with threats that can be asymmetric, where nation-state and non-nation-state actors share the same contested space, the United States will need vision and authority, and it will need to implement an effective strategy or agency that involves multiple stakeholders, both public and private, in government, military, academia, and the commercial sector. The US does not have one group or organization to address the breadth of issues that we face as a nation. We are on an unacceptable path, allowing our national interest to drift with no one at the helm – these issues require vision and leadership.

Building a 'federal cyber board'

Today, there are far too few calling for the kind of national leadership we need for cyberspace. Some acknowledge the security threat and have begged for legislation to enhance the collective security of our nation. Some have rightly fought for issues that ensure freedom in cyberspace. Unfortunately, the national dialogue is not addressing the right challenges; it should be focused on the collective breadth of cyberspace, including its opportunity as well as its risk. The Honorable James M. Simon, in 'Toward a General Understanding of Cyber Safety' (2012) summarizes the need in three parts:

- Prevent all users, including the least aware, from becoming a threat to the larger community.
- Influence and change laws by gaining cooperation from companies even as we continue to encourage growth and creativity.
- Cooperate with other governments, and create an international coalition of the willing.

There is a need for the creation of an organization singularly focused on cyber across all its many facets – the technical, the risks, the challenges, and the opportunities. For such an organization to be successful, it must blend the strengths of the public and private sectors for cyber. Further, it needs to be designed with insulation from the political (recognizing a purely apolitical organization would be impossible). Modeled on some of the characteristics of our Federal Reserve, a 'Federal Cyber Board' could represent a promising alternative. The name 'Federal Cyber Board' seems inappropriate, but so do Cyber Agency, Cyber Administration, Cyber Federation, Cyber Bureau, Cyber Activity, or Cyber Department; so rather than distract with the name, we'll call it the FCB and focus on how it will accomplish the goal. This FCB should have a charter to ensure free and open access to cyberspace (the connected digital world), enable the conditions for growth in and through cyberspace, create a shared threat awareness, and form a common security partnership.

In the beginning, a government and private-sector partnership must be in the DNA of this new FCB organization. The cyber domain is dominated by the private sector, which has the best talent, agility, and awareness. However, commercial and private entities do not have the charter (nor the skill) to look out for the breadth of US national interests. That charter belongs to the federal government. Merging the equities of numerous stakeholders in cyber will not be easy, but our nation needs the best of both worlds. This new FCB will thrive if it is designed with a partnership and teamwork ethos.

A fundamental element to 'designing partnership in' for the FCB is to create it with a leadership team – a diverse board of governors from the key stakeholder communities. With 12 governors, we can ensure the right equities are represented. Thus, three governors should be individuals associated with large cyber 'market-maker' companies such as Apple, Google, Cisco, Microsoft, Facebook, and Symantec. An additional three governors need to represent the entrepreneur perspective in cyber. Plausibly these governors would come from individuals with a successful entrepreneurial past, such as founders of new firms that achieved market success in cyber-security applications. Further, three governors need to be representatives of our critical infrastructure such as public utilities, telecommunications infrastructure, and the financial sector.

Finally, there will be a need for the expertise of those who understand serving our nation and its broad interests. The last three governors should come from a pool of diplomatic and security leaders with a charter to represent the nation's interests in the security and international engagement arenas. The idea is that the best candidates would come from Department of Defense, Department of Homeland Security, Department of State, and academia. By appointing governors for six-year terms and staggering the terms, the FCB can elevate above short-term political noise and focus on the longer, more important mandate to enhance our nation's competitiveness and security in the cyber arena. Like the Federal Reserve, presidential nomination and Senate confirmation for FCB governors also helps ensure a process to deliver the right leadership.

This FCB, led by such a board of governors, would be the premiere entity to drive our national strategy in cyberspace. Governors would uniquely sit at the forefront of technology and policy. The FCB chair should be expected to regularly brief Congressional committees. Governors should meet with members of Congress as well as leaders in government agencies, academia, the military, cyber advocacy groups, and the breadth of cyber companies. Their vision should be to ensure the open flow of communication while providing oversight and creating sustainable partnerships.

The board of governors represents four major constituencies. Those entities outline the first order of organization for the four branches of the FCB:

- A branch focused on ensuring and reinforcing success where it already exists
- A branch focused on fostering innovation and entrepreneurship
- A branch focused on security
- A branch focused on policy and engagement

Similarly to the Federal Reserve, which has been given significant national power to issue national currency and to supervise our monetary system (with an ability to respond effectively in stress/crisis), the FCB should be charged with an overall mission of improving our nation's economy, competitiveness, and freedom of access, even as it ensures the cyber security of the nation.

Again, similar to the Federal Reserve, the FCB ought to find a significant level of autonomy. While this might be the most contentious element of an FCB, it will probably be the secret to its success. Political stresses tend to be focused on the short term. Cyberspace demands a longer vision and grander strategy. With governors appointed by the president and accountability to consult with Congress, the FCB would be involved in the government, but it would be independent of both Congress and the administration. To ensure success, the powers delineated to the FCB ought to be adequate to ensure the FCB can actually execute its mission. Powers delegated to the FCB need not eliminate the ability of either the executive or legislative branches of government to act on cyber issues. Legislation still provides the best avenue to execute national-level decisions, and in fact, the FCB should seek to be the advocate for the right cyber legislation. Further, federal cyber legislation can and should be responsive in a critical situation similar to legislative/executive actions to pass fiscal stimulus laws in response to the economic crisis of 2008.

As we develop the depth of the organization, the structure of the FCB must also reflect how our nation operates in the cyber world. The private sector must have a significant stake in the execution of FCB policy/planning/operations. Not only does the private sector own much of the critical services that must be protected, but they also set the market that will influence future cyber growth. Motivated by innovation, expansion, creativity, and profit, those in this private-sector space have a unique vantage within the cyber arena, and arguably, they comprehend it best. Therefore, a significant portion of the FCB staff should derive from the workforce that was developed in the private sector. All four FCB branches need this private-sector expertise.

In the FCB security branch, there must be a blend of private and public. Recognizing the FCB will be held responsible for security of something it does not own, the personnel should leverage its private-sector ethos and the expertise of government security. The private sector usually leads innovation and owns the vast majority of the nation's critical infrastructure and cyber networks – a significant challenge for an FCB tasked to defend and respond to serious cyber threats to our nation while at the same time empowering opportunity and growth. When we look at the traditional idea of national security, we find cyber security, inside and outside the government, as a crucial element. According to Etzioni (2011),

> Protecting the private sector is increasingly critical because the US, more than most, if not all, other nations, draws heavily on private corporations for ensuring national security. Corporations manufacture most of the nation's arms. Corporations produce most of the software and hardware for the computers the government uses. And corporations, under contract with the government, carry out many critical security functions, including the collection and processing of intelligence and the conduct of covert operations.

This concept of national security in cyberspace (for an FCB to protect) is further expanded when we consider how our society relies on sectors outside of government for vital activities we consider essential for our society and thus for our security, especially for the finance industry, the electrical grid, and our energy industry. Needless to say, the mission of an FCB will demand a new approach to security – one where government and the private sector partner far more than we have seen in the past.

Tasks for an FCB

Prior to her departure from the Department of Homeland Security, Janet Napolitano stated that

> [c]yber security is a shared responsibility [and] emerging cyber threats require the engagement of our entire society. The success of our efforts to reduce cyber security risks depend on effective communication

> and partnership among departments and agencies from all levels of government, the private sector, international agencies and the American public (Napolitano 2012).

While acknowledging her intent, it is clear an FCB would be far better constructed to handle this challenge than the current structure of our federal government. Undoubtedly, we need public-sector guidance for our national security. Equally important, however, we must ensure the FCB, as a quasi-government entity, will not act and become directive in a manner that unnecessarily stymies our national development, cooperation, and innovation. As James Lewis eloquently stated, '[W]e need to find a balance between the requirements of public safety, security and growth. Too much regulation will kill economic growth, too little will put the country at risk' (Lewis 2014).

Foundation efforts

A competitive nation needs to prepare its workforce for the future. Our society is changing, and a cyber-skilled workforce will be a key to setting our path to development. The FCB should tackle this critical need by bridging industry's needs with our academic communities. This is more than just developing programmers. As every element of our economy touches cyber, we need workers able to leverage all its power. For example, autoworkers need the ability to work with robots, bankers need to work with a variety of mobile and dispersed communication devices, and police need to leverage big data for investigations. This workforce development will require attention from elementary school through universities, from technical schools to online continuing education.

Our cyber infrastructure, while held in private hands, is a national asset. Fiber backbones, connectivity across national boundaries and electromagnetic spectrum, are the first places we see a need for a comprehensive national strategy to ensure we are able to capitalize on the power of cyber. On a related path, our nation needs robust and energetic research and development (R&D) efforts for cyberspace. The FCB can provide the catalyst for difficult projects, assist in removing impediments to private R&D, and offer recommendations to Congress for national-level investment.

Policy

According to Clemente (2013), several policymakers do not grasp the importance of the role played by the private sector in the cyber ecosystem. Most policies addressing cyber-security challenges in Western countries are frequently contested by national and international entities. It is only very recently that national governments came to the realization that their strategy to secure the cyberspace is inadequate and needs to be renewed. A FCB configured to be the partnership of private and public sectors should be prepared to respond to this dilemma. It should provide the wide perspective of US citizen opinion and thought combined with an understanding of the technology to enable wiser decision-making.

As the United States tackles difficult societal issues, an FCB could provide the nucleus of expertise and advocacy, encouraging an open cyber world that enables free-market ideas and expansion, promoting free trade, focusing investment in cyber, thinking strategically, preparing for connectivity needs, and ensuring freedom/privacy. None of these issues is easy, but an FCB provides a focal point for tackling them with logic and with a collaborative mindset.

Standards: the right approach

The FCB can tackle the tough balancing act of helping with cyber security without becoming restrictive by regulation. It is important to keep in mind that companies have a fiduciary responsibility to their shareholders and stakeholders, both internal and external. To shift corporate incentives more heavily toward cyber protection of US citizens does not lie within their authorities, nor, for the most part, their interests. In some cases, corporate executives speculate about how much security is enough and wonder if growing security costs will negatively impact profit margins and, subsequently, shareholder confidence. Furthermore, regulatory pressure threatens to alienate trusted partners. Information sharing and recommendations for cyber-security standards have not been uniformly adopted, but an FCB with clear authority to lead in this space can help mitigate the already demonstrated danger of fragmentation and gaps in both growth and protection. In seeking solutions, it is also clear that a 'one size fits all' approach is not the answer. Today's National Institute of Standards and Technology Framework for Improving Critical Infrastructure Cyber Security takes the right approach. With the FCB in charge of this framework, we will gain a cyber-focused organization with a leadership role in measuring success, applying best practices and improvements, sharing information, allocating resources correctly, and creating a sustainable vision for the future in cyber.

Cyber security

Cyber security is crucial to the success of our nation. Today's cyber-security environment is dynamic but diffused. Three major activities are key to security and resilience: prevention, detection, and responding to an attack. An FCB can be a leading catalyst for each of these three activities.

First, the FCB can become the convening entity for collaborative preventative cyber security using the free market to deliver collective security to the broad cyber user community at minimal cost. Users need a 'place to go' in cyberspace to validate their devices are 'clean' from known malware. The FCB can be that place. By centrally convening access to all cyber-security firms in one location, users would be able to access the breadth of cyber-security expertise. If necessary, they would then have the right connection to a security company to clean their devices, if infected. Cyber-security companies would benefit by being part of the collective that generates a huge volume of activity. Users would benefit by routinely getting the 'Good Housekeeping Seal of Approval' of being clean. Everyone can benefit, for example users get relatively inexpensive cleaning, security companies get huge volume, commerce sees improved security across the spectrum, and our society is less at risk. This is the virtual equivalent of flu shots that inoculate a large portion of society and help prevent the outbreak of a pandemic.

The FCB can also be the collaborative point for detection of cyber attacks. First, the FCB can help identify and endorse the right kind of technologies to improve our ability to determine malicious activity in real time. Further, by establishing a robust information-sharing regime, the FCB can help to rapidly identify the vectors of an attack sequence. This arena is one where the FCB can help facilitate the success of other government agencies (DHS, DOD, DOJ) and expand the scope to our broader commercial sector.

Finally, the FCB can provide the foundation for response to malicious cyber activity. Through the use of standards, proliferation of best practices, and convening exercises, the FCB will be able to assist at all levels – from the individual user to corporate levels to a full-scale national response.

International engagement/leadership

Since cyberspace is a domain that transcends national boundaries, our nation must address our global equities. Engaging in the international forums and in bilateral venues, the FCB can assist our government agencies with responsibilities in this arena. At the core, the FCB can be the supreme advocate of freedom in cyberspace. By recognizing a future that supports a globally connected world, the FCB could harness relationships with partners as well as developing and emerging economies, work to sustain future harmonies among universal partners in order to collaborate on opportunities, minimize risk, support mutually benefiting economic development, and promote appropriate technology transfer.

Conclusion

The immediate and future challenges with and in cyberspace demand audacious action. Cyberspace offers immense promise to improve our economy, our productivity, and our lives. On the other hand, there are very real, costly, sophisticated, and pervasive threats that jeopardize not only individuals, but could undermine the fabric of our economic and national security. Our risk grows daily as we become increasingly reliant on cyber. The US needs an entity to lead and focus on cyber across all its many facets – the technical, the risks, the challenges, and the opportunities. In order to help make that possible, the answer should come through the creation of a new cyber agency, one poised to address a wider range of needs. An FCB is the answer to tackle the challenges while capturing the strengths of the private sector and providing national strategic vision.

To say there are important resistances for another government entity would be an understatement. Administrations from both parties have made their fair share of effort to maneuver inside the complexities of our interagency authorities and, at the same time, attempt to shape the cyber future of our nation. Unfortunately, as of today, those attempts have fallen quite short.

9

Defending Critical Infrastructures Against Cyber Attacks: Cooperation through Data-Exchange Infrastructure and Advanced Data Analytics

Frederic Lemieux

Introduction

In the United States government, executive and legislative branches have taken several steps to outline a new initiative for cyber security and information sharing. For instance, section four of the *Executive Order – Improving Critical Infrastructure Cyber Security* (2013) stipulates that 'it is the policy of the United States Government to increase the volume, timeliness, and quality of cyber security information shared with US private sector entities so that these entities may better protect and defend themselves against cyber threats.'[1] More precisely, the federal government, through the Department of Justice (DoJ), Department of Homeland Security (DHS), and the Director of National Intelligence (DNI), has issued directives to promote the timely production and release of unclassified reports on cyber threats to the United States. Also, the executive order calls for the development of a Cyber Security Framework, which should include 'a set of standards, methodologies, procedures, and processes that align policy, business, and technological approaches to address cyber risks.'[2] As a part of the Cyber Security Framework, DHS is required to establish a voluntary program to support the adoption of the framework by owners and operators of critical infrastructure. DHS has also been mandated to establish a set of incentives in order to promote participation of the private sector in the voluntary program.

This chapter proposes an approach to create a trustworthy cyber-defense information-exchange platform that will enhance cooperation within and across the critical infrastructure sectors in sharing cyber-security information. This data-exchange platform pursues three main goals: (1) developing a data-exchange platform housing cyber-security data in a trustworthy, cooperative environment; (2) resolving existing limitations related to law and policy, security, and automation of exchanges in cyber security through inception of advanced technology that enhance information-sharing confidence; and (3) creating an incentivizing data-exchange process by providing advanced analytical capabilities and information access that will support operational and strategic decision-making of participants. The innovative nature of this research project resides in the development of new methods and approaches related to information sharing and information management. First, the creation and integration of an agile data model capable of handling independent topic ontologies to data-sharing platform is certainly a genuine contribution to the field of information management.

This chapter is divided in three sections. Section one addresses the fundamental challenges posed by information sharing. Section two outlines a review of existing initiatives on cyber-security information exchanges and identifies limitations that impinge on the capacity to share information effectively. Section three illustrates in detail the nature of the proposed information-exchange platform, the articulation of the proof of concept, and the characteristics of advanced analytical capabilities to be deployed. Finally, this chapter concludes on the broader applications of the cyber-security information-exchange system.

Fundamental challenges posed by information sharing

Despite the basic incentive of exchanging information to improve the tracking, correlating, and disseminating of information for greater awareness, there are several limitations that continuously undermine information sharing on cyber incidents (inspired from the work of Sheptycki 2004). First, the *non-reporting and non-recording* of information is, without a doubt, the less technical and the more human or organizational behavior problem of all. In general, both organizations and individuals do not report or record information because of a simple cost/benefits calculus in which the cost related to the information-sharing process is always superior to its benefits (time, new resources, new technology, security dispositive, etc.). Usually information sharing will occur in a more compliant environment rather than a voluntary one.

Another challenge is the *digital divide* between participants. For instance, information storage and operating systems are different, and some sectors are highly reluctant to incept changes because of the legacy of system (Seacord, Plakosh and Lewis 2003). There is also a *duplication of data-exchange systems* that competes against each other or duplicates the collection and storage of data, thereby increasing information gaps and linkage blindness. Duplication of data-exchange systems causes deterioration of information sharing and encourages information hoarding (Arenas et al. 2013). *Linkage blindness* occurs when information is not shared due to a lack of exchange standards between the participants. This problem is not a technical one like the digital divide but is more of a system problem for which participants have to define a horizontal framework to support information exchange (Egger 1984). The *information gap* is often a direct consequence of the previous limitations but can also result from hierarchical information systems or from a discrepancy in sharing processes between several participants (Ramakrishnan and Gehrke 2002).

Noise is mainly generated by an overload of information transiting in a communication channel between a transmitter and receiver. Noisy channels of communication are counterproductive and impact the capacity to identify the relevant information due to interference often induced by unstructured information. Information can be filtered to ensure that organizations receive the information that is most relevant to their successful management of an emergency (Mendonca, Jefferson and Harrald 2007).

Information overload is more than noise and happens also when participants do not have the capability to structure, analyze, and visualize shared data. This problem often leads to a misuse or underuse of information and contributes to sharing data without a precise purpose (Eppler and Mengis 2003, 2004). One main explanation for information overload is the *compulsive data demand* that occurs when more emphasis is on the volume of data demanded instead of focusing on the collection of better data. The critical problem resides in the quality versus quantity of data that information systems are integrating and which are directly affecting the reliability of the data (Phythian 2013).

Another important challenge is *institutional friction,* which can directly affect information sharing due to competitive or conflicting interest among participants. This friction can be attenuated if participants, who may be competitors, can cooperate with each other to reach a higher value creation if compared to the value created without interaction and the struggle to achieve competitive advantage (Asaro 2011; Nash 1944).

The concept of *coopetition* is certainly a central element of an effective cyber data-exchange framework, since participants from the same critical infrastructure sector have to work together for the collection and sharing of information about cyber threats but at the same time must compete for product and service market share.

So far, several organizations and security information-exchange communities have worked together to find solutions to information-sharing limitations. The past ten years have been characterized by a rapid growth of data-sharing mechanisms and standards. The next section outlines some of the most prevalent information-sharing systems, protocols, and standards across industries.

Existing information-sharing mechanisms and standards

Currently, several information-exchange systems related to cyber threats are not entirely automated and depend mainly on human intervention through e-mail, telephone calls, and face-to-face meetings. This section provides an overview of the existing framework related to information exchanges about cyber incidents. More precisely, we will explore sharing mechanisms that have an international and national outreach, such as Cyber Security Information Exchange Framework (CYBEX), European Information Sharing and Alerting (EISAS), Collective Information Framework (CIF), information sharing and formatting standards developed by Mitre (TAXII, STIX, CyBox), information sharing and formatting standards developed by Internet Engineering Task Force (IODEF, RID), and Financial Sector Information Sharing Analysis Center (FS-ISAC). The section also provides characteristics of data-exchange systems and their capabilities. This review of existing information mechanisms and standards is certainly not exhaustive, but it provides the opportunity to identify the inherent strengths and weaknesses, providing critical insight to articulate the parameters and requirements of an integrated approach for cyber-threat data sharing.

Cyber Security Information Exchange Framework (CYBEX)

In 2010, the United Nations, through its agency the International Telecommunication Union (ITU), launched the first iteration of standardization for the development of CYBEX. The goal of this approach is to provide a common format and framework for assured information exchange on a global scale and to minimize the disparity of cyber-security information availability between organizations located in different countries. CYBEX also aims at facilitating information exchange through standardized techniques. This framework operates on five functional blocks: (1) the information description block that structures cyber-security information for exchange purposes (which includes the Security Content Automation Protocol developed by the US National Institute of Standards and Technology); (2) the information discovery block that identifies and discovers cyber-security information and entities; (3) the information query block that requests and responds with cyber-security information; (4) the information assurance block that ensures the validity of the information, and (5) the information transport block that exchanges cyber-security information over networks (Rutkowski et al. 2010).

The CYBEX model is articulated around six structured information clusters (Adegbite et al. 2010). The first cluster focuses on exchanges related to (a) knowledge, such as platforms, weaknesses, vulnerability, and exposure; and (b) state, such as security state measurement, configuration checklists, and assessment results. The second cluster addresses exchanges related to incident and investigation, such as event expression, incident patterns, malware patterns, and malicious behavior. The third cluster concentrates on exchanges related to information-sharing policies. The fourth cluster is composed of identification, discovery, and query mechanisms. The fifth cluster addresses identity assurance, such as trusted platforms and authentication assurance methods and levels. The last cluster is focused on exchange protocol, such as trusted network connection, interaction security, and transport security. Though CYBEX focuses on information exchange, it does not provide solutions on how to generate the information that is to be exchanged (for example, 'traceback' technology) and how to use the exchanged information effectively.

European Information Sharing and Alerting System (EISAS)

The EISAS is a new concept developed by the European Network and Information Security Agency (ENISA) and supported by the European Commission.[3] The main goal of the EISAS would be to raise awareness about information technology security issues among citizens as well as small and medium enterprises across Europe. In 2012, the ENISA developed a pilot of the European Information Sharing and Alerting System, which was deployed in six member states. The large-scale deployment pilot focused on two main aspects: collaboration among identified key players and sharing as well as distributing good practice information (Birkas and Bourgue 2013). The pilot was conducted in an experimental setting, meaning that the shared content was limited, focused, and pre-produced. The role of the participating disseminators only extended to a one-time translation and dissemination through their direct outreach channels.

The pilot has generated several findings about the feasibility of a full implementation. The most important result was showing that 'participants willing to provide information have to be supported by some entity that takes over the task of information post-processing (from the point of view of the information provider) and the task of information pre-processing (from the point of view of the information disseminator)' (Birkas and Bourgue 2013: 4). Moreover, the experiment provided insights related to management and administration of the system. For instance, the necessity of providing advice to the information providers on how to produce distributable information; the need to incept an incentive system that encourages the information producer to participate; the necessity of assisting the information provider and the information consumer with professional services related to information translation, information localization, technical troubleshooting, and dissemination of best practices. In other words, the overall outcome of the pilot shows that if cyber-threat information sharing runs without support, it is highly unlikely, at least in the early phases, to establish foundational community building.

Collective Information Framework (CIF)

The collective information framework has been jointly developed by the Research and Education Networking, Information Sharing Center (REN-SAC) in collaboration with Indiana University, Internet, and National Science Foundation. The CIF helps organizations to structure, normalize, store, post process, query, share, and produce data sets related to cyber-security threats in a single database. More precisely, this intelligence management system combines known malicious threat information from several sources and uses that information for incident response, detection, and mitigation. The majority of the information contained in the CIF repository includes IP addresses, domains, and URLs associated with malicious activity.[4]

According to Kijewski and Pawlinski (2013: 7–8), the data model used by CIF is based on the Incident Object Data Exchange Format. In order to enable efficient queries, selected metadata is extracted and put in dedicated tables that serve as indexes referencing back to complete incident data. The intelligence management system runs a set of data enrichment routines on the newly collected events. The system has client-server architecture with a single server that is entirely dedicated to storage and processing. The system periodically generates feeds of recent reports for every type of threat based on the means that can be used to identify that particular threat. Queries are performed through the CIF client application and constitute the only means of distributing/accessing data.

Trusted Automated Exchange of Indicator Information (TAXII™), Structured Language for Cyber Threat Intelligence Information (STIX™), Cyber Observable Expression (CyBox™)

Jointly developed by the Department of Homeland Security and MITRE in 2012, the *Trusted Automated Exchange of Indicator Information* is a standard that supports automated sharing, confidentiality, and authentication. It aims at facilitating the exchange of structured cyber-threat information (Mitre 2013a). More precisely, the TAXII provides a set of specifications defining the network-level activity of the exchange and defines services as well as messages to exchange data. It supports multiple sharing models including variations of 'hub and spoke' as well as 'peer to peer.' The TAXII enables organizations to share the information they choose about cyber threats with the partners they choose. The *Trusted Automated Exchange of Indicator Information* is not a specific information-sharing initiative or application and does not attempt to define trust agreements, governance, or other non-technical aspects of cyber-threat information sharing. The TAXII is supported by other critical initiatives such as *Structured Language for Cyber Threat Intelligence Information* (STIX) and *Cyber Observable Expression* (CybOx).

The CybOX and STIX are standards that support data formatting. The STIX language intends to convey the full range of potential cyber-threat information. The STIX architecture consists of eight constructs and 70 object types that capture all the details relating to a malicious actor's campaign, tactics, actions, and targets (Mitre 2013b). These constructs are (1) threat actor; (2) campaign; (3) trusted third party; (4) exploit target; (5) courses of action; (6) incident; (7) indicator; and (8) observable. CybOx is a standardized schema for the specification, capture, characterization, and communication of incidents that are observable in the operational domain (Mitre 2013c). It can provide information on event management/logging, malware characterization, intrusion detection, incident response/management, attack pattern characterization, and geolocation information.

Incident Object Description Exchange Format (IODEF) and Real-time Inter-network Defense (RID)

The Incident Object Description Exchange Format (IODEF) was developed in 2007 by the Internet Engineering Task Force, and it is a standard for representing computer security information commonly exchanged between Computer Emergency Readiness Teams (CERTs) and organizations facing a cyber-security incident. According to Danyliw et al. (2007: 3), the main objective of IODEF is to 'provide an improved ability to resolve incidents and convey situational awareness by simplifying collaboration and data sharing.' More precisely, the format increases automation in processing of incident data, decreases effort in normalizing similar data from different sources, and provides a common format on which to build interoperable tools for incident handling and subsequent analysis. An interesting particularity of IODEF format is that it can be easily compatible with STIX format by defining an extension of the STIX incident base type which leverages IODEF for representing the incident information. The architectural alignment provided by STIX's incident type extension would be lost, but this extension would allow integrating IODEF incidents within the broader STIX framework (Mitre 2013b).

The IODEF format can be supported by *Real-time Inter-network Defense* (RID) protocol as a report message (Moriarty 2010). According to Moriarty and Trammell (2010: 2) 'RID outlines a proactive inter-network communication method to facilitate sharing incident handling data while integrating existing detection, tracing, source identification, and mitigation mechanisms for a complete incident handling solution.' The IODEF and RID are information-sharing standards in the Multi-national Alliance for Collaborative Cyber Situational Awareness (MACCSA) information-sharing framework.

Financial Sector Information Sharing Analysis Center (FS-ISAC)

The FS-ISAC was incepted in 1999 by the financial services sector in response to Presidential Directive 63 on Critical Infrastructure Protection: Sector Coordinators (1998).[5] The center serves as a forum composed of hundreds of members that share information about cyber threats. The FS-ISAC is capable of instantly and globally sharing critical and authoritative information across the finance industry. The center provides anonymous information-sharing capability and professional services from industry experts that verify and analyze cyber-security threats, identify solutions, and formulate recommendations to be disseminated to the FS-ISAC members.[6] The center uses information-sharing and formatting standards such as TAXII and STIX.

The rationale behind the FS-ISAC is to develop a trusted forum to exchange information about physical and cyber threats posing serious risk to participants in the finance industry. In such context, organizations have decided to form a club for which members have to pay a fee to access and share critical information, notifications, and more specialized services such as incident response procedures and prevention solutions. This information-exchange forum aims at facilitating cooperation between competitors in order to create added values, such as rapid dissemination of critical information about cyber threats, analysis, and adapted solutions. However, a recent study has pointed out 'that many information sharing alliances' membership policies are plagued with the incentive misalignment issue and may result in 'free-riding' or 'no information sharing equilibrium' (Liu, Zafar and Au 2014: 3). In their study, the authors present a simulation model in which they have integrated a new membership policy including an insurance option. The result demonstrates that the new policy is in alignment with members' incentives and leads to a socially optimal outcome. Other policy options can also be developed to decrease the proportion of 'free-riders' and improve the information-sharing equilibrium.

This section proposed a review of a selected list of cyber-threat information-sharing mechanisms and standards used in the United States and globally. The field of cyber defense and information sharing is evolving rapidly, and several other exchange mechanisms and standards have been developed or will soon be available. Nonetheless, several limitations continue to persist and reduce the capacity of organizations to effectively and securely share information about cyber threats. Also, some other relevant dimensions of limitations have received far less attention, notably those related to deciding on what to share, with whom, when, as well as reasoning about and adapting to the potential repercussions of sharing.

In a recent publication, Dandurand and Serrano (2013) identify a series of existing challenges that need to be addressed: (1) lack of automation for large-scale information sharing; (2) inconsistencies and errors in existing data and information quality assurance varies; (3) limited accessibility to some data sets stored on the Internet or pertaining to commercial products such as anti-viruses; (4) lack of interoperability between exchange mechanisms and protocols due to proprietary rights; (5) incompatible semantic used to label data; (6) limited support to foster collaboration efforts and mobilization of subject matter experts that can assist with information exchange; and (7) mistrust and skepticism regarding the security and regulation mechanisms related to the dissemination of shared information. In the following section, we will present the design of the proposed approach and how this research project can integrate existing information-sharing mechanisms and protocols while addressing most challenges currently encountered by existing frameworks.

Proposed approach

The field of security information exchange has greatly evolved in the past ten years, but several technical and behavioral problems need to be addressed. The effective protection of critical infrastructure against cyber threat requires an information-exchange structure that is agile, resilient, and secure in order to improve cyber defense strategies and tactics. This chapter proposes a research and development (R&D) framework that aims at designing a new information-sharing approach that fulfills these generic requirements through the conceptualization and operationalization (proof of concept) of a cyber-security data-exchange infrastructure.

More precisely, we are proposing the creation of a data-exchange infrastructure that will support the cyber-security ecosystem by allowing a secure integration of existing information-sharing communities and resilience through decentralization of data storage. One of the main outcomes of this project is to support automated information exchanges through different knowledge markets (open and private) and allow organizations to publish and subscribe to several data offerings. This new approach also addresses the information-sharing limitations presented in previous sections. The R&D framework enables a market-driven approach to the development of cyber-security information infrastructure and can encompass all sectors of critical infrastructure identified by the Presidential Policy Directive 21 – Critical Infrastructure Security and Resilience (2013).

Architecture design

This system builds on an existing concept of information-exchange systems, but the proposed framework integrates new capabilities that provide a unique infrastructure in the field of cyber defense. The main requirements of the proposed cyber-security information-exchange system are derived from the work of North Atlantic Treaty Organization (NATO), which has defined 11 high-level requirements in consultation with officials from more than 20 countries (including the United States). These consultations have been conducted with the objective of conceptualizing a cyber-security data exchange infrastructure known as CDXI (see Dandurand and Serrano 2013). Exploiting conceptualization of the cyber-security data-exchange infrastructure is appropriate to this project because the complexity of exchanging cyber-security information with NATO's members is reflected in the requirements and applies to the complex reality of information sharing between the owners of critical infrastructures in the US. Those requirements stipulate that the information-exchange infrastructure should do the following:

1. Provide an adaptable, scalable, secure, and decentralized infrastructure (e.g., allow for flexible access controls, private query, anonymous contribution, data replacement, data exchange review, support of small and large organizations, and knowledge exchange among community of interest).
2. Provide for the controlled evolution of the syntax and semantic of multiple independent data models and their correlation (e.g., allow organizations to implement standardized data models that can be captured by 'independent topic ontologies' through data correlations).
3. Securely store both private and shared data (e.g., allow organizations to use agile data models and correlation capabilities to discriminate private and sharable data).
4. Provide for customizable, controlled multilateral sharing (e.g., allow integration of IPEs from all participating organizations and enable free association of IPEs with communication channels).
5. Enable the exchange of data across non-connected domains (e.g., allow exchange across these 'air gaps' without compromising secured networks).
6. Provide human and machine interface (e.g., allow for the integration and use of graphical user interfaces as well as application programing interfaces).
7. Provide collaboration tools that enable burden sharing for generation, refinement, and vetting of data (e.g., allow organizations to collaborate through threaded discussion mechanisms in order to annotate data and understand context of data modification).
8. Provide customizable quality control process (e.g., allow organizations to share quality control processes in order to ensure collaboration in maintaining data quality assurance).
9. Expose dissension to reach consensus (e.g., allow organizations to correct detected errors and inexactitudes by integrating divergent values that expose multiple truths).
10. Support continuous availability of data (e.g., allow organizations to store and access data locally in case of interruption or disconnection of external networks – resilience).
11. Enable commercial activities (e.g., allow integration of accounting models for data providers and data consumers through commercial contracts).

Agile data model and ontology

The management of large volumes of information demands technology that automates the acquisition process as much as possible, which requires an agile data model that can integrate different types of data models and ontologies. The cyber-security data-exchange infrastructure will also support automated correlation between data offerings as well as across knowledge markets. Through the application programing interface, the automation will also allow the activation of alert systems, semi-automated response systems, and fully automated response systems. To achieve this level of automation, some ontological problems related to data offerings will need to be resolved. One of the several challenges of this research project is to move from a specific ontology model adopted by organizations to an 'independent topic ontology' model (ITO) that can recognize specific data models and allow correlation with other data models (relationship between attributes, taxonomy, entities, etc.). Table 9.1 below provides cyber-security information (attributes) to be shared by the participants throughout the testing phase of the project.

The extraction and processing of languages or element specifications in cyber security require a flexible data-management approach that allows the integration of new or existing standardized data models. Therefore, the agile data model must also permit a learning ontology process in which new supervised ontological knowledge can be produced from the introduction of new languages or specifications. The supervised learning ontology process can be achieved through the application of several techniques, such as statistical and syntactic analysis (Downey et al. 2004),

Table 9.1 Cyber-security data attributes

Incident ID	Attack Type
Alternative ID	Attack Patterns and Vectors
Related Activity	Target Type
Detect Time	Software (Applications and Operating Systems)
Start Time	Patches and Fixes
End Time	Protocol Specifications
Report Time	Certifications
Description	Additional Data (verification, remediation, patch, fixes)
Assessment	Event Data (event report)
Method (vulnerability, weakness, platform ID)	History (frequency)
Contact	

pattern-based extraction (Ruiz-Cassado et al. 2007), and clustering (Cimiano et al. 2005). This feature offers the opportunity to considerably improve data offerings from publishers.

Enforcement of information-exchange policy

Another critical challenge is to integrate/automate Information Exchange Policy (IEP) developed by participating organizations and pair them with communication channels between participants. The IEPs provided by organizations will also provide all information necessary to specify encryption and authentication protocols, authorization mechanism, and preferred ontology. The cyber-security data-exchange infrastructure should allow the activation of relevant information exchange policy based on the originators' and recipients' requirements. The exchange of cyber-security data will require translation of requirements documented (e.g., legislation, policy, MOUs, SLAs, etc.) into sets of system-enforceable rules that are traceable and auditable throughout their life cycle. This procedure has been applied to some extent in the healthcare domain by enforcing information-exchange policy and requirements on organizations sharing personal information about patients (Kuperman 2011; Marchibroda 2007; Vest and Gamm 2010).

Also, the data center industry is using architecture that automates a virtual infrastructure while automatically applying relevant security mechanisms associated with information-exchange policy (Cabuk et al. 2010). These applications are interesting because they provide examples of some challenges related to the release of information based on levels of trust and operational context; translating legislative mandates, policy, and information-sharing agreements; and transforming information-exchange policies into system-enforceable rules. The cyber-security data-exchange infrastructure will support automated correlation between data offerings as well as across knowledge markets. Through the application programing interface, the automation will permit automatic activation of alert systems, semi-automated response systems, and fully automated response systems.

Privacy, security, and trustworthiness of data

In an information-exchange infrastructure, protection of privacy and security of data are paramount not only to foster confidence in the reliability of the exchange system but also to increase trust among participants. The conceptualization of the proposed cyber-security data-exchange infrastructure offers several security challenges, and several layers of protection must be addressed. Techniques for fine-grained and context-based access control are critical to protect the confidentiality and privacy of the data against insider threats (Bertino 2014). In addition to the insider threat and data privacy, there is the fundamental issue of trustworthiness of what is communicated. Shared cyber-security data can be contaminated by malicious activity and contain erroneous information. Establishing trust requires at least the possibility of authentication of transmissions and establishing trust in their contents.

The proposed cyber-security data-exchange infrastructure will allow high-level encryption and authentication protocols to be integrated. Also, it is acknowledged that data trustworthiness is an issue greater than security, and the proposed data-exchange infrastructure will allow solutions combining different approaches and techniques including data integrity, data quality, record linkage (Inan et al. 2012), and data provenance (Sultana, Shehab and Bertino 2013). For example, the proposed cyber-security data-exchange infrastructure should allow integration of tools that enable the refinement and vetting of data. Moreover, participants should be allowed to work collaboratively and reach consensus on information dissension by disclosing divergent values to information managers.

Another important technique is the anonymization of data that provides the participants a channel in which they can engage in anonymous communication and avoid the consequences of identity revelation. There are several reasons why a participant may prefer sharing information anonymously, such as reputation degradation, adverse financial repercussions, and legal liabilities. For example, studies have demonstrated the potential ripple effect of sharing security information that can result in loss of market share and stock market value due to negative publicity (Cavusoglu et al. 2004; Campbell et al. 2003). In a recent study, Dunning and Kresman (2013) have developed an algorithm for anonymous sharing of private data among several participants that assign anonymous identities, which are unknown to other members. This algorithm provides high-level resistance against collusion among other participants. Moreover, according to the authors, 'the assignment of serial numbers allows more complex data to be shared and has applications to other problems in privacy preserving data mining, collision avoidance in communications and distributed database access' (Dunning and Kresman 2013: 402). Similar types of algorithm should be employed in this project to provide a higher level of participant identity protection and to increase the likelihood of participation in cyber-security data exchange.

Governance and operating rules

The information sharing is based on a 'collective governance' approach that brings various stakeholders together to engage in consensus-oriented decision-making (Ansell and Gash 2007). They will also develop a guide that defines the governance, technical elements, and practices related to the use of this cyber-security data exchange system. This guide will be largely inspired by existing rules and procedures already established by sharing communities, such as the financial and energy sectors (FS-ISAC and ES-ISAC). Finally, operating rules must address aspects related to knowledge markets and data usage accounting and integrated to a module within the cyber-security data-exchange system.

Information sharing incentive and advanced analytics

Effective and flowing exchange of information within an environment that is not under any legal or administrative compliance requirement can be extremely difficult to achieve. In a voluntary information-sharing market, participants often need strong and meaningful incentives to open the communication gates and expose sensitive information to other organizations. In all domains, in addition to incentives, information sharing and cooperation between individuals and organizations requires a certain level of trust that most participants will reciprocate. If a participant believes that leeching is a dominant behavior in the network, the motivation to share information will rapidly decline. To address the issue of leeching, some research has proposed a sanction mechanism based on an evolving genetic algorithm to enhance trust between several nodes transmitting data packets (Seredynski et al. 2007). Also, other research has employed the game theory and applied the prisoner's dilemma approach to randomly selected persistent players (Cascella 2008). The study showed that by incepting a reputation system allowing punishment, it was possible to distinguish good players (information sharers) from bad players (leechers). However, in a voluntary information-sharing market, the use of punishment systems would have very limited consequences.

Nonetheless, some researches have shown that encouraging good behaviors instead of punishing poor performances can increase the likelihood of participants increasing their involvement toward a common goal. Although observing high levels of cooperative behavior is not enough to sustain information sharing in the long run, participants that receive social approval can have a significant positive impact on cooperative behavior (Cheshire 2012). The study conducted by Li (2011) shows that the perceived value of sharing information is associated with the opportunity to receive an incentive for exchanging information. Among all variables included in the analysis, social approval was the strongest predictor of the willingness to share information. The benefit for and interest in the community are also important predictors.

Conversely, the cost of sharing information was not significantly related to the willingness to contribute. This is not surprising, since one of the most cited reasons for joining an information-sharing network is the cost savings generated by the decrease of security breaches (see Table 9.2). Reducing the financial impact of cyber-security attacks represents, in many cases, the main goal of alliances such as Financial Sector-Information Sharing Analysis Center (FS-ISAC). A survey conducted by the European Union Agency for Network and Information Security (ENISA 2010), in collaboration with several large organizations sharing cyber-security information, demonstrate that economic incentives stemming from cost savings is the most important motivation to share information. The answers from the participants suggest that cost savings are generated by quicker reactions to threats, vulnerabilities, and attacks, or from anticipating network failures. For example, FS-ISAC has been able to inform their members of key information rapidly enough for them to avoid large-scale denial of service attacks. The publication of such success in preventing attacks generated an increase in the interest of joining information-sharing alliances and increased contribution of existing members. However, if publicizing success can demonstrate the value of sharing information, it also carries risk related to negative perception, such as creating the false perception of collusion within the industry.

Another critical incentive motivating organizations to share security information is having the access to high-quality information that is available in a timely fashion and specific enough to serve decision-making processes or operational purposes. It is also important that participants share information that has equivalent value and equal relevance to participants' concerns. The study of Gal-Or and Ghose (2005: 187) shows also that an increase in cyber-security data sharing may lead to two benefits for organizations: 'a direct effect, which increases demand, and a strategic effect, which alleviates price competition. These incentives become stronger with increases in the firm size and the degree of competition.' Other noticeable incentives identified by the ENISA survey include the existence of established trust among participants, receiving privileged information from government agencies, incentives deriving from the process and structure of sharing, and fostering participants' autonomy while ensuring company buy-in.

As mentioned previously, the main incentives in sharing information come from the cost savings associated with prevention and mitigation of cyber attacks. Also, the quality and pertinence of the information shared has an important value to the participating organization in the sense that it becomes highly applicable to operational and strategic needs. Using analytical capability to generate collaboration between organizations and to increase information sharing has been proven successful. For example, the work of Lemieux (2010) and Gerspacher and Lemieux (2010) on Europol showed that when the agency decided to transfer analytical capabilities through training and expertise, the level of criminal intelligence shared among member states increased drastically, even for the most reluctant countries. Because national police agencies understood the operational and strategic values of using criminal intelligence from neighboring countries, they started to contribute to Europol Information System (EIS) on a more regular basis while continuing to develop analysis capabilities in collaboration with Europol.

The integration of data analytics capabilities to the cyber-security data-exchange infrastructure can certainly provide an added value in terms of generating incentives to participate and contribute to the information sharing. More precisely, participants choosing a passive

Table 9.2 Ranking of incentives to information sharing (Source: ENISA 2010: 16)

High	Medium	Low
1. Economic incentives stemming from cost savings	3. The presence of trust among IE participants	7. Economic incentives from the provision of subsidies
2. Incentives stemming from the quality, value, and use of information shared	4. Incentives from receiving privileged information from government or security services	8. Economic incentives stemming from gaining voice and influence
	5. Incentives deriving from theprocesses and structures for sharing	9. Economic incentives stemming from the use of cyber insurance
	6. Allowing IE participants' autonomy but ensuring company buy-in	10. Incentives stemming from the reputational benefits of participation
		11. Incentives from receiving the benefits of expert analysis, advice, and knowledge
		12. Incentives stemming from participants' personal preferences, values, and attitudes

role would have access to query/search functions through structured query language (SQL) to find and correlate information shared by participants. Participants choosing to play a more active role will have access to a more sophisticated cyber-security data analytics toolkit. This analytical capability will help data analysts as well as data managers to better understand the massive, multidimensional data sets to protect their organizations. Also, the toolkit will provide the technique and the technology to perform data mining and analysis for both operational and strategic purposes. For example, in addition to data visualization, analysts and data managers will be able to perform analyses related to risk assessment, trend analysis, pattern recognition (event and behavioral), cyber attack modeling and simulation, impact assessment, response evaluation, and so on. These analytical techniques will allow participants to produce internal reports covering several purposes: incidents detection, warning notification, and incident response.

Finally, to address the importance of data quality as incentives for participants (e.g., relevance, accuracy, timeliness, comparability, coherence, and clarity), the cyber-security data exchange infrastructure will incorporate tools and techniques based on sound methodology, appropriate statistical procedures, non-excessive burdens on participants, and cost effectiveness (Batini et al. 2009). Approaches such as process variables, quality indicators, quality reports, user surveys, self-assessment, and peer-review audits can be developed and implemented to assure the quality of data exchanged in the infrastructure (Ehlin and Korner 2007). Moreover, having organizations involved in the quality assurance process can foster cooperation among participants and increase the level of trust in the infrastructure, which is critical to information sharing.

Conclusion

The application of the proposed project goes above and beyond the simple exchange of cyber-security data. The development of an agile data model and independent topic ontologies offer the unique opportunity to circumvent traditional limitations related to interoperability of information sharing's standards and protocols. Moreover, the agile data model embedded within a data-sharing system helps to transcend technological constraints of older information systems. This exchange structure can allow recent or old information systems to share data no matter what type of topic ontologies the organizations employ. Such a data-exchange framework can allow diverse organizations pertaining to different sectors of the critical infrastructure to communicate sensitive information about cyber incidents or threats. Moreover, the system can allow owners of critical infrastructure to share information in compliance with security and privacy rules through the automated enforcement of information-exchange policies. These exchanges can also occur between US and non-US organizations within or across critical infrastructure sectors (e.g., exchange among NATO members) that want to expand their understanding of the scope and depth of cyber-security incidents and cyber threats in other countries. Finally, the proposed data-exchange infrastructure can be relevant to other fields, such as law enforcement and the sharing of criminal intelligence between federal agencies as well as state and local police forces. This exchanged data can include structured and unstructured information, such as audio and video files. Again, the agile data model, the independent topic ontologies, and the automated enforcement of security and privacy policies can drastically improve the interoperability of the multiple information systems already existing.

Notes

1 http://www.whitehouse.gov/the-press-office/2013/02/12/executive-order-improving-critical-infrastructure-cyber security. Accessed on 12 June 2014.
2 Ibid. Previously cited.
3 European Commission Communication on Critical Information Infrastructure Protection – COM (2009)1493
4 https://code.google.com/p/collective-intelligence-framework/wiki/WhatisCIF Accessed on 12 May 2014
5 http://www.gpo.gov/fdsys/granule/FR-1998-08-05/98-20865 Accessed on 18 June 2014.
6 https://www.fsisac.com/about

10

Cyber Resilience: A Review of Critical National Infrastructure and Cyber-Security Protection Measures Applied in the UK and USA[1]

Wayne Harrop and Ashley Matteson

Introduction

Imagine a future world held to ransom by the demands of a hidden unnamed force lurking behind computer screens and operating in murky shadows. Imagine the same forces causing indiscriminate and lasting harm in an ever-increasing technologically dependent world. A world where a private GPS or home smart meters could be hacked and reprogrammed remotely without the owner's permission or knowledge; where financial systems might unexpectedly suffer malicious downtime; where ICT systems are hacked to steal highly confidential information or intellectual property; where power grids are interfered with, water treatment plants remotely breached and attacked by digital terrorists and cyber activists, and public transportation networks (ICT systems) targeted to cause maximum chaos during peak travel periods. Maybe this all sounds like another Hollywood blockbuster movie. However, consider the alternative that it is a real prospect. If so, what can be done to safeguard the basic way of life? In particular, what is being done by countries like the UK and USA to safeguard the security people have come to expect and enjoy but may take for granted?

The nature of cyberspace and everyone's growing reliance upon it is constantly changing and the way advanced users operate in a modern, decentralized cyberspace environment provides good cover and anonymity for an intelligent foe, making the attribution of any cyber attack very difficult to pinpoint. One thing is certain, cyber attacks are growing at an alarming rate worldwide, and this includes both the UK and the USA. The threat is especially focused and targeted toward government systems, business, and commerce. The public might get caught up in the crossfire, where the threat infiltrates what they rely upon to sustain their daily activities. Recent examples include attacks on Barclays and Santander banks. 'Barclays bank [was attacked] using a remotely-controlled KVM (keyboard-video-mouse) device and 3G routers. [The attack was] described by police in the United Kingdom as being a 'Mr Big' of UK cybercrime. [The attackers] are said to have stolen £1.3 million ($2 million) from Barclays bank, before being caught' (Dunn 2013). In addition to the Barclays attack, Santander also recently foiled a cyber plot of a similar nature. Whatever the motive, cyber security is a very hot topic. There are several critical questions that must be answered in relation to cyber security to ensure the protection and integrity of IT systems and data. How prepared are the UK and USA for sustained and targeted attacks on their respective essential services delivered to the public through critical national infrastructure (CNI) and critical information infrastructures? Who is taking the lead on protecting national security on the public's behalf?

One certainty exists, cyber-security concerns are becoming more apparent every day, and the issue is likely to grow as a real and present challenge to the smooth functioning of any modern Western economy. The USA recognizes cyberspace as a fifth domain of its own national security agenda in tandem with pre-existing domains such as land, sea, air, and space. As such, the United States established the United States Cyber Command (US CYBERCOM) in 2009 to recognize that fact and organize a body under the US Department of Defense (DoD) to address cyber issues. On 22 February 2013, at the Fourth Annual Cyber Security Conference in Washington, DC, United States Air Force Major General Brett T. Williams, Director of Operations at US Cyber Command: said 'part of Cybercom's mission is to help in defending the homeland, especially against cyber attacks and other activities in cyberspace that could affect national security' (Williams 2013). There are strong and compelling reasons why it is important to protect CNI from cyber attacks, but there is also an ill-defined enemy behind the emerging trend of cyber attacks. The enemy could vary, and understanding their evolving capabilities and organizational limits is crucial to fending off cyber attacks orchestrated by a range of possible foes, such as state-sponsored attackers, hackers, anarchists, and criminal gangs. Recently, former Secretary of Defense Leon Panetta in the USA stated:

> A cyber attack perpetrated by nation states or violent extremist groups could be as destructive as the terrorist attack of 9/11, such a destructive cyber terrorist attack could paralyze the nation (Garamone 2012).

The DoD is believed to be probed millions of times a day by malicious cyber actors. By September 2011, DoD had identified over 70 million cumulative malware threats against its own networks. In the last few years, malicious actors have launched cyber attacks against America's nuclear infrastructure, advanced military weapons systems, water treatment facilities, credit card companies, financial institutions, and the NASDAQ stock exchange (United States of America National Senate, Democratic Policy and Communications Center 2012).

The Internet of things

The world is entering a new future reality where nearly anything that can be on the Internet will be:

> Internet of Things (IoT) is an integrated part of Future Internet including existing and evolving Internet and network developments and could be conceptually defined as a dynamic global network infrastructure with self-configuring capabilities based on standard and interoperable communication protocols where physical and virtual 'things' have identities, physical attributes, and virtual personalities, use intelligent interfaces, and are seamlessly integrated into the information network (Vermesan et al. 2009).

With the growth of the IoT, consumers should be cognizant that even on a local level there is already an expansion of the national infrastructure reaching into every future home, which may be connected to smart-enabled devices, sensors, and wireless devices essentially controlled by remote systems in support of compelling reasons, such as the use of household energy consumption monitoring by service providers, the networking of household smart appliances, and improved access to remote digital content (such as cloud and streaming media content).

> Many people think of the Internet of Things as some magic web of connected devices that will communicate with each other and act together, but the reality is probably closer to the vertical segmentation we already have in our lives. So while we might have an Internet

> of electricity that combines elements of the smart grid with our thermostats, we may have to buy some kind of device that we plug our appliances into to connect them to the Internet of electricity (Higginbotham 2013).

The additional devices might be simple routers or other networking equipment that will connect the devices together in the home so they can communicate with each other as well as providing the streaming services, such as weather forecasting and Internet radio broadcasts, to the user. Although IoT brings much convenience to the consumer, it could also introduce great vulnerability and must be a part of future cyber-security strategies. As many of the IoT devices are used to manage power usage, they often feed the smart utility grids operated by power companies. As such, the overall approach to securing these new IoT devices against cyber attack is by protecting the smart grids they are connected to as well as employing basic Internet security in the home networks they are connected to, such as firewalls and routers that only allow traffic to the connected devices that are expected (streaming radio feeds, weather forecasts, outgoing data stream to the smart grid reporting usage). Much of the areas discussed above have some direct or indirect relationship to CNI.

The approach to cyber security in the UK

In the UK, the Centre for the Protection of Critical National Infrastructure (CPNI) defines CNI as follows:

> Those facilities, systems, sites and networks necessary for the delivery of the essential services upon which daily life in the United Kingdom depends and which ensure the country continues to function socially and economically (Centre for the Protection of National Infrastructure 2013a).

There is a compelling need to better understand, protect, and maintain critical assets and information infrastructures against cyber threats, especially in a world where 80 percent of private-sector industries operate national assets as part of their core business. There is limited consumer and end-user understanding or technical skills to counter the growing cyber threats. In many cases in the corporate world, a weakness is clearly found where organizations have porous controls around intrusion detection and monitoring, incident response, or computing forensics. Cyber issues can, of course, be as a result of internal and external sources to any organization or system, requiring organizations to look within as well as to the exterior interface with the world at large. As organizations look both internally and externally, they are better able to create a robust cyber-security posture to combat would-be attackers from whatever vector the attackers may originate.

An excellent example of a threat presented by an external source but accepted or introduced from an internal source is a successful spear-phishing attempt. Recently (as of September 2013), the CPNI released new guidance to educate the public on this topic called 'Spear Phishing: Understanding the Threat.' The CPNI guidance explains that spear phishing is 'a targeted form of email deception that results in exploitation or compromise of individual devices and organizational networks' (Centre for the Protection of National Infrastructure 2013b). The CPNI guidance explains how spear-phishing attacks work, the likelihood of being targeted, and the steps an organization can take to manage the business risks. In addition, the European Union Agency for Network and Information Security (ENISA) has identified and published the first comprehensive cyber threat landscape analysis of 2012, summarizing over 120 threat reports. The report lists the top ten threats identified by ENISA and their trends and concludes that drive-by exploits in emerging technology areas have become the top web threat. The areas considered by ENISA when generating the list include mobile computing, social media/technology, critical infrastructure, trust infrastructures, the cloud, and Big Data. The identified top ten threats mentioned above are as follows:

1. Drive-by exploits (malicious code injects to exploit web browser vulnerabilities)
2. Worms/trojans
3. Code-injection attacks
4. Exploit kits (ready-to-use software package to automate cyber crime)
5. Botnets (hijacked computers that are remotely controlled)
6. (Distributed) Denial of Service attacks (DDoS/DoS)
7. Phishing (fraud e-mails and websites)
8. Compromising confidential information (data breaches)
9. Rogueware/scareware
10. Spam

(European Union Agency for Network and Information Security 2013)

The issue of cyber security is so relevant and topical for the British government that it developed and published the National Security Strategy in 2010, which essentially describes how – 'in an age of uncertainty – the

UK needs the structures in place to allow it to react quickly and effectively to new and evolving threats to UK security.' The National Security Strategy identifies 15 priority risk types, and one of the top four risks identified includes the need to safeguard against 'hostile attacks upon UK cyber space' in line with national emergencies, such as a serious pandemic flu outbreak (Centre for the Protection of National Infrastructure 2013a).

The UK and the US are responding to an ever-shifting landscape and engineered and well-thought-out cyber-attack capabilities. Organizations are waking up to the growing calls to stem the consequences of cyber attacks, but they also need to make the right decisions and understand the trade-off between performance, cost, and risk as a sustainable business model. Effective resilience requires an understanding and broader atonement to infrastructure assurance within organizations and the future direction the organization and its competitors and customers are moving in. The modern organization will need to be able to better anticipate and forecast cyber risks and vulnerabilities connected to new and emerging ICT trends, such as the explosion in smartphone usage and the forces shaping the digitalization of commerce and society, and connect this understanding to the security investments and planning of expenditures over asset life cycles.

In response to the rise in cyber threats, a new Government Communication Headquarters (GCHQ) Security Operations Centre was established in 2009 within the UK and declared in full operation as of March 2010. Its mission is to provide real-time opportunity for businesses and organizations to report the cyber attacks they are experiencing in the hopes to gain assistance in stopping the attacks and learning from them to prevent future attacks. A BBC news report of this open exchange between government and industry says: '[T]his should give the government early warning of cyber attacks that could bring down critical national infrastructure. In return, the commercial sector can expect expertise on tap' (Centre for the Protection of National Infrastructure UK Government 2013b).

In addition to GCHQ's Security Operations Centre, Francis Maude, minister for the Cabinet Office, in March 2013 developed a new Cyber Information Sharing Partnership (CISP), previously called Project Auburn. CISP started in February 2011, when the prime minister met with the 'captains of industry' to discuss cyber security and attacks. Both industry and government agreed that faster situational awareness was required in light of the severity and rapidly increasing pace of cyber attacks against UK interests and industry. At the CISP launch, Frances Maude stated:

> We know that cyber attacks are happening on an industrial scale and businesses are by far the biggest victims of cyber crime in terms of industrial espionage and intellectual property theft with losses to the UK economy running into the billions of pounds annually.
>
> [The CISP will] introduce a secure virtual 'collaboration environment,' where government and industry partners can exchange information on threats and vulnerabilities in real time. The CISP will be complemented by a 'fusion cell,' which will be supported on the government side by the Security Service, GCHQ and the National Crime Agency and by industry analysts from a variety of sectors. They will work together to produce an enhanced picture of cyber threats facing the UK for the benefit of all partners (European Union Agency for Network and Information Security 2013).

CISP is being driven principally by the Centre for the Protection of National Infrastructure (CPNI), the Department for Business, Innovation and Skills (BIS), and GCHQ. The Cyber Intelligence Fusion Cell will promote information sharing between industry sectors and enrich intelligence using multiple sources. The Fusion Cell will work closely with a number of partners from defense and finance sectors supported by a few UK government agencies with a stake in national cyber security.

The Fusion Cell will monitor cyberspace via a giant screen showing where in the UK cyber attacks by foreign states and criminals are emerging. The information will be shared among up to 160 top British companies under the CISP. The Fusion Cell will comprise about ten officers from MI5, GCHQ, and MI6, as well as handpicked specialists from some of Britain's biggest companies (UK Government 2013a).

The UK government is responding because cyber security is such a serious issue, where cyber-related fraud and intellectual property theft alone is estimated to cost the UK's economy £27bn per year. This mid-range financial estimate was identified by the UK government in 2011 and then represented a breakdown of £21bn of costs to businesses, £2.2bn to government, and £3.1bn to citizens (Watts 2011). In contrast, a report by the National Counterintelligence Executive in the USA has also described a persistent, widespread campaign by foreign nation-states to steal intellectual property and trade secrets from American companies.

'Chinese actors,' it found, 'are the world's most active and persistent perpetrators of economic espionage' (Garamone 2012). A recent study conducted by Norton, an Internet security company, estimates that, during a year, cyber crimes – including identity theft and online scams – cost the USA $140bn in cash and lost time. It found the $388bn global cost of cyber crime to be greater than the black market for marijuana, cocaine, and heroin combined (UK Government 2013b). Although, traditionally, the finger has tended to point toward the Chinese government or rogue groups in China, the scale of the problem is growing elsewhere in the world with reports of rising cyber threats and capabilities emerging from India, the Middle East, and Eastern Europe.

American private security firm Mandiant, based in Virginia, recently identified the headquarters of Unit 61398, a People's Liberation Army (PLA) group suspected of waging cyber warfare. The study revealed that 150 highly sophisticated cyber attacks against targets in the US had originated from inside. Unit 61398 looks like any other 12-storey tower on the outskirts of Shanghai's Pudong (Sky News 2013). Governments are aware of a number of hacking hotspots around the globe but are often reluctant to openly point the finger at countries, mainly for diplomatic and trade reasons.

Nonetheless, the hidden scale of cyber-related crime is said to be enormous, and it is growing significantly, especially where organizations are reluctant to discuss their status as a target by a stream of well-networked, savvy opponents who are able to collectively pool ideas and design, customize, and deploy an array of increasingly cunning and resourceful methods through cyberspace. A drive for better digital literacy in children will bring both opportunities and threats to the future of cyberspace interactions. Educating the end-user seems sensible, but the rate at which technology and software changes makes it hard to maintain.

ICT investments are increasingly driving the delivery and monitoring of critical infrastructure with networked apparatus, and this leaves the door wide open to cyber attacks, such as phishing, man in the middle browser attacks, malware, Trojans, worms, root-kits, distributed denial of service (DDoS), and increasingly evolved and well-planned no-notice attacks, such as 'zero day' attacks, one of the biggest emerging concerns (British Broadcasting Corporation News 2011). In the face of such security challenges, organizations are scrambling to improve cyber incident management and intrusion detection and deploy plausible decoys such as honey nets and honeypots. Absolute reliance on the constant availability of CNI to fuel basic life needs presents a serious and increasing challenge that requires better stakeholder coordination and improved understanding of this fast-moving, ill-defined problem.

The CPNI is one source that provides detailed public guidance aimed at advising organizations in the UK on how to better understand and manage their own current cyber-security arrangements. The CPNI recommends a total of 20 specific controls (with sub-controls) spanning across various technical measures and activities (Symantec Corp. 2011), with the primary goal of helping UK organizations prioritize their efforts to defend against the current most common and damaging computer and network attacks (see Table 10.1).

The need for clear guidance is most relevant where organizations are scrambling to improve their own cyber situational awareness. One crucial area not evident in the CPNI's 20 controls listed in Table 10.1 (and seldom in the minds of many company executives) is the need for integrated business continuity arrangements, specifically addressing the resilience and backup arrangements for critical ICT infrastructure and information resources. This is especially important if critical ICT systems are compromised and taken offline by a determined attacker. Clearly, conducting a business impact analysis as part of a continuity process should add value to cyber priorities by informing on 'defined criticality' and 'recovery times' for any organization's critical infrastructure. Further to the above, linking cyber resilience into the organization's (and its critical suppliers' and contractors') polices and strategic risk registers will place the issue at the heart of governance procedures and firmly across a broad range of stakeholder agendas.

In addition to the guidance to industry from the CPNI, the British government is also behind the Communications Electronic Security Group

Table 10.1 CPNI cyber-security guidance aimed at organizations in the UK

Critical control 1	Inventory of authorised and unauthorised devices
Critical control 2	Inventory of authorised and unauthorised software
Critical control 3	Secure configurations for hardware and software
Critical control 4	Continuous vulnerability assessment and remediation
Critical control 5	Mahvare defences
Critical control 6	Application software security
Critical control 7	Wireless device control
Critical control 8	Data recovery capability
Critical control 9	Security skills assessment and appropriate training to fill gaps
Critical control 10	Secure configurations for network devices
Critical control 11	Limitation and control of network ports, protocols and services
Critical control 12	Controlled use of administrative privileges
Critical control 13	Boundary defence
Critical control 14	Maintenance, monitoring and analysis of security audit logs
Critical control 15	Controlled access based on the need to know
Critical control 16	Account monitoring and control
Critical control 17	Data loss prevention
Critical control 18	Incident response capability
Critical control 19	Secure network engineering
Critical control 20	Penetration tests and red team exercises

(CESG), a cyber incident response scheme launched in November 2012, which provides access to companies certified to respond to the consequences of cyber attacks. This scheme builds upon the Cabinet Office '10 Steps to Cyber Security,' which was launched in September 2012. It is aimed at business leaders, describing the cyber-security threat and providing advice on the basic measures to increase cyber security within their organizations. CESG's aim is to 'protect the vital interests of the UK by providing policy and assistance on the security of communications and electronic data, working in partnership with industry and academia' (Harris 2013).

Hacking and cyber attacks against the defense sector is particularly concerning, so the UK government established the Defence Cyber Protection Partnership (DCPP), which aims to meet the emerging threat to the UK defense supply chain by increasing awareness of cyber risks, sharing threat intelligence, and defining risk-driven approaches to applying cyber-security standards. The DCPP currently partners the CPNI, GCHQ, the Ministry of Defence, and nine companies: BAE Systems, BT, Cassidian, CGI, Hewlett Packard, Lockheed Martin, Rolls-Royce, Selex ES, and Thales UK (Ponemon Institute 2012).

Lessons from cyber attacks

From the serious attacks on the Estonia government network, it is already known that disruption to national infrastructure and vital resources can have a profound and cascading impact, seriously challenging essential and basic public services. Maintaining the safe and efficient use of Internet-enabled and networked services such as communications, energy, finance, food, government, health, transport, and water supplies presents governments and stakeholders with a stark challenge in the face of increasing levels of sophisticated cyber attackers and hackers. For the most part, CNI is tightly coupled and connected, with little or no slack (designated redundancy), and it is spread across complex geo-spatial and multidimensional boundaries with critical node points or hierarchical controlling systems, such as SCADA systems, across the UK. The threats posed to national interests have brought stakeholders such as the Metropolitan Police, GCHQ, CPNI, and the Cabinet Office together to ensure cyber initiatives are coordinated in line with national risk assessments and a National Risk Register.

The approach to cyber security in the USA

After the 2007 DDoS on Estonia and other cyber attacks that followed, the president of the United States, then President George W. Bush, knew that action was needed to secure America's critical infrastructure against cyber attack. In January 2008, President Bush signed the National Security Presidential Directive 54/Homeland Security Presidential Directive 23, Comprehensive National Cyber Security Initiative (CNCI), which was initially classified until March 2010, when President Barak Obama released public information of the CNCI and its main recommendations. 'President Obama has identified cyber security as one of the most serious economic and national security challenges we face as a nation, but one that we as a government or as a country are not adequately prepared to counter' (Centre for the Protection of National Infrastructure 2013c). President Obama used the CNCI as part of an in-depth cyberspace policy review that he commissioned to assess America's readiness to withstand cyber attacks. When the review had concluded, President Obama reviewed the results and released a plan using the CNCI goals to secure America's digital infrastructure.

The main goals of the CNCI are the following (Centre for the Protection of National Infrastructure 2013c):

- To establish a frontline of defense against today's immediate threats by creating or enhancing shared situational awareness of network vulnerabilities, threats, and events within the federal government – and ultimately with state, local, and tribal governments and private-sector partners – and the ability to act quickly to reduce current vulnerabilities and prevent intrusions
- To defend against the full spectrum of threats by enhancing US counterintelligence capabilities and increasing the security of the supply chain for important information technologies
- To strengthen the future cyber-security environment by expanding cyber education, coordinating and redirecting research and development efforts across the federal government, and working to define and develop strategies to deter hostile or malicious activity in cyberspace

In addition to CNCI, the USA has continued its recognition that cyber security is a growing challenge to its own national security by creating the US CYBERCOM, which was mentioned briefly earlier. US CYBERCOM is one way that America is fighting cyber terrorism, both in the public (United States military) and private (home and business) sectors.

In addition to the work done each day by US CYBERCOM, the United States also recognizes that emergency response and readiness teams are needed to assist when an attack takes place, or in advance of it to stop or mitigate the potential effects of the attack. As such, the United States

Computer Emergency Readiness Team (US CERT) maintains a website to help the less technically savvy as well as expert information technology professionals. 'US CERT's mission is to improve the nation's cyber security posture, coordinate cyber information sharing, and proactively manage cyber risks to the nation while protecting the constitutional rights of Americans' (CESG 2013). The US CERT provides users with specific instructions on how to protect their computer systems and networks from attack by making readers aware of the newest patches available by software vendors that should be installed to protect any information systems that access the Internet. In addition, US CERT provides a large library of articles and discussions on cyber-security-related content to educate readers on the newest threats and related trends and how to defend against them. They also give users helpful tips offering best practices and advice on security issues of interest to the general public. Their last and likely most recognized role is to provide security alerts and vulnerability bulletins, also providing viewers with the link to download the patches to mitigate the vulnerability.

While US CERT, US CYBERCOM, and the CNCI have been in existence for a few years, the job of cyber security is far from finished. President Barrack Obama recently signed a presidental executive order requiring the United States to step up its cyber security to better combat the potential for attacks on critical infrastructure. The Improving Critical Infrastructure Cyber Security Executive Order (EO) was signed on 12 February 2013. 'The EO tasks the National Institute of Standards and Technology (NIST), within the Department of Commerce, to develop a baseline Cyber Security Framework that sector-specific agencies would rely upon to establish a voluntary critical infrastructure cyber security program' (Government of UK 2013b).

Cyber threats: Weapons of mass disruption

The modern computer virus is rapidly evolving and becoming capable of causing massive disruption to critical infrastructure and vital resources. Project Aurora in the United States was a controlled demonstration on how a virtual breach afforded control over a critical industrial control system. The system was tampered with in such a way that it caused rapid physical malfunction and failure to occur despite fairly robust security measures being in place. The fear is that a sustained and well-targeted campaign of cyber attacks could easily cause mass disruption to daily functioning across government, industry, and commerce (as reflected in Estonia to a lesser degree).

The impact of the Stuxnet virus on industrial control systems

Maintaining effective cyber-security capabilities in complex distributed infrastructures requires extended vigilance and situational awareness in a changing cyber landscape. Stuxnet introduced a need to change the way that organizations and anti-virus providers managed threats from cyberspace. Stuxnet had a very effective strategy for the covert monitoring of specific targeted facilities:

> The United States of America, Department of Homeland Security said that this 'highly complex' computer worm was the 'first of its kind.' Stuxnet's potential to damage CNI caused 'worldwide alarm,' according to the *Financial Times*, and has been called a 'paradigm shift' by the European Network and Information Security Agency (US National Security Council).

Stuxnet deployed a sophisticated code that had an intelligent control interface, allowing for extensive and refined damage capabilities in networked components. Iran was reportedly the intended target due to its own controversial nuclear development programme. Symantec Corporation stated that 'the Iranian organisations were involved in "normal" industrial projects. These conclusions are based on intercepted data that Stuxnet transmitted to its command and control server' (US National Security Council). Analysts suggest that Stuxnet was specifically designed to target the reliability of Siemens components, controlling the rate at which nuclear centrifuges could safely spin in Iranian nuclear plants. 'Tehran confirmed in September 2010 that Stuxnet had infected about 30,000 IP addresses in Iran. The high infection rate increases the probability that Iranian centrifuge facilities may have been affected, but is not in itself proof that they were' (US National Security Council).

The impact of the Flame virus on critical information infrastructure

Flame was detected in May 2012 and it was believed to essentially exhibit cyber-espionage capabilities. The virus was believed to have been around for two years prior to its discovery. The Middle East, including Israel, Syria, and Iran became particularly vulnerable. Flame had the ability to remotely control and pass information from webcams. It could take and remotely send screenshots from infected computers and control microphones, switching them on and off as required in any infected computing devices. In addition to webcam control, it can also record all network connections on the infected machine, gather basic systems data, search and steal files based on name or contextual window searches, and it can scan for and use locally connected Bluetooth devices. Flame allegedly had 20 times the code of Stuxnet according to Russian anti-virus provider Kaspersky.

The International Telecommunications Union of the United Nations said:

> Flame is a suite of tools for professional cyber-espionage. It is an example of powerful cyber weapons that are a rising international problem. People's lives could be seriously affected by such weapons if personal data is stolen, or if critical infrastructure is threatened through connections to the Internet (US CERT 2013).

The threats are getting worse each day, so the defenses must keep pace with them with constant virus definition updates and patches. This is why keeping on top of vulnerability patching of personal and business computers and devices should be top of every Internet user's mind. Cyber threats exist because there is a time gap between recognizing the changing nature of the threat landscape and concerted action to limit the impacts of a potential targeted attack.

The USA and the UK's classification of CNI

Both the UK and USA face similar cyber threats at home and from foreign sources. There are, however, some crucial differences observed around what is considered critical infrastructure and, secondly, which agencies at a national level have an active responsibility for nationwide cyber security. Much of these differences are reflected in the legal processes and structural design and functioning of each country's own government.

In 2008, the Department of Homeland Security identified 17 critical infrastructure and key resources (CIKR) sectors. These CIKR sectors were recognized under the United States National Infrastructure Protection Plan, and CIKR were mapped against 15 existing US emergency support functions (Daubert et al. 2013).

Since 2008, the Presidential Policy Directive 21 (PPD-21): Critical Infrastructure Security and Resilience has advanced to include a national policy to strengthen and maintain secure, functioning, and resilient critical infrastructure. This directive now supersedes the pre-existing Homeland Security Presidential Directive 7. PPD-21 now identifies 16 critical infrastructure sectors, as shown in Table 10.2 (Barzashka 2013).

Table 10.2 The 16 CIKR sectors in the USA

Chemical sector
Communications sector
Dams sector
Emergency services sector
Financial services sector
Government facilities sector
Information technology sector
Transportation systems sector
Commercial facilities sector
Critical manufacturing sector
Defence industrial base sector
Energy sector
Food and agriculture sector
Healthcare and public health sector
Nuclear reactors, materials and waste sector
Water and wastewater systems sector

Table 10.3 The nine CNI sectors in the UK

Communications
Emergency services
Energy
Financial services
Food
Government
Health
Transpon
Water

The UK addresses the scope of critical infrastructure and applies a different approach (although there is much overlap) to the USA. In essence, the UK's national infrastructure is currently categorized into only nine sectors by the CPNI (see Table 10.3).

There are some cross-sector themes, such as technology, wherein there may be infrastructure that supports the delivery of essential services across a number of sectors. The Cabinet Office in the UK has in place a National Cyber Security Strategy. This area has four main strategic objectives:

- Making the UK one of the most secure places in the world to do business in cyberspace
- Making the UK more resilient to cyber attack and better able to protect its interests in cyberspace
- Helping shape an open, vibrant, and stable cyberspace that supports open societies

- Building the UK's cyber-security knowledge, skills, and capability

A significant proportion of funding (£650m over five years) has already been given to organizations such as GCHQ by the British government to improve the detection of cyber attacks on the UK's interests. The funding will help transform the UK's situational awareness in cyberspace. A series of investments will see GCHQ and partners further increase the ability to respond to a diversified range of cyber threats and to protect the UK's national and economic security interests (International Telecommunications Union of the United Nations 2012).

The United States' approach has been similar in using agencies such as GCHQ to begin to confront cyber-related challenges. The National Security Agency and the Department of Homeland Security work very closely together to fight cyber attacks, especially when they are directed toward those critical infrastructure areas defined in PPD-21 and other related legislation. Presidents George Bush and later Barack Obama have placed a high priority on securing cyberspace so it is safe for all Americans and world users. Every year, for the past three years, some form of cyber-security bill has been introduced, and every year for the past three years civil liberties organizations such as the American Civil Liberties Union and the Electronic Frontier Foundation have fought hard for privacy (Federal Emergency Management Agency 2008).

More recently, a legislative bill was introduced in the USA called the Cyber Security Act 2012 by Senators Joseph Lieberman (I–CT) and Susan Collins (R–ME) that would have allowed for better critical infrastructure protection, but at the cost of privacy rights. The bill, if passed, would have allowed for collection and monitoring of any electronic transmission that contained certain keywords related to planning to cause harm to the United States' critical infrastructure. The proposed Cyber Security Act of 2012 was defeated in the US Senate. There were serious concerns about public privacy rights, and it was felt by some detractors to be too restrictive and burdensome on businesses (Jaycox 2012). The proposed act would have allowed private corporations to voluntarily share suspicious online activities with the intelligence and law enforcement communities. Despite the Cyber Security Act 2012 not being passed into law and to ensure that the critical infrastructures were still protected from cyber attack, in February 2013, President Obama signed an executive order titled Improving Critical Infrastructure Cyber Security. It appears the debate on cyber security will continue as the rate of cyber intrusion, disruption, espionage, and destructive attacks will increase worldwide.

The release of sensitive information about the National Security Agency's clandestine mass surveillance programme called 'PRISM' has added another dimension to cyber-security protection measure discussions and divided public opinion about electronic surveillance and the security of cyberspace. The release of restricted information by Edward Snowden to the *Washington Post* and *Guardian* newspapers has caused great debate about civil rights, checks and balances, and approaches used by the US and UK intelligence communities in pursuit of keeping their own citizens safe from harm. The dimensions of cyber security, and who is sourcing information on whom, will become far more complex and interesting as time goes by, as will the way US and UK government leaders balance personal liberties and human rights with national security.

Conclusion

This chapter has identified how critical infrastructure is pivotal to the smooth running of daily life. As everyone continues to become more dependent on critical infrastructure in a growing cyber-enabled era, the threats and vulnerabilities are likely to grow and change faster than people can perhaps appreciate or collectively respond. The nature of surveillance for national security is likely to also stray into the debate with civil libertarians.

The perpetrators behind foreign cyber attacks are not always easy to determine, and the motives behind attacks can vary significantly. As attack surfaces (e.g., via mobile-enabled devices such as 'bring your own device') change, there is a continued requirement for much better situational awareness and planning at all levels, ranging from home personal computer users to corporate enterprises. The nature of the threats ranging from disruptive, criminal, and destructive cyber attacks requires greater research and cooperation between industry, commerce, infrastructure owners, infrastructure operators, and government(s). This process of partnership has begun to take shape in the UK and US with developments such as the US' Comprehensive National Cyber Security Initiative (CNCI) and UK's Cyber Information Sharing Partnership (CISP) at nationally coordinated levels.

The approaches and methods used to manage cyber threats at a national level in the UK and US differ most notably in defining the scope of each country's respective CNI sectors. Government agencies in both the US and UK have acknowledged the issues concerning cyber threats, but there is still much more work to be done, and the success of any strategy is dependent upon applying effective foresight and controls based on understanding the landscape of cyber and future developments (Cabinet Office, Government of UK 2012).

Additionally, when comparing the UK and US approaches, it is evident that both countries recognize clearly that their critical infrastructure is vulnerable and that the government must assist in protecting against cyber attacks. Both countries recognize that standards are needed to help businesses and organizations to adapt their processes to accommodate a security posture that would better protect them from cyber attacks. Legislation is another option (such as the recent failed Cyber Security Bill 2012), but this is not without considerable opposition, emanating most notably from civil liberties groups (Jaycox 2012). It is possible that the capabilities of a defensive posture against aggressive cyber attacks will move toward a more offensive posture, involving attacking the adversary under clear standing orders, being empowered to execute those orders immediately upon discovery of the threat or in anticipation of it, thereby making the strategy more effectively aligned to cyber warfare than cyber resilience. It is also clear that most nations engage in some form of heightened surveillance and cyber espionage, whether they admit it or not.

Note

1 This chapter has been previously published in the *Journal of Business Continuity and Emergency Planning* (2014), 7(2): 149–62.

Section IV

Cyber Intelligence and Information Security

11
Typologies of Attacks and Vulnerabilities Related to the National Critical Infrastructure

Charles Pak

Introduction

Cyber infrastructure provides a backbone for economic stability, growth, agility, and new business opportunities. Cyber vulnerabilities arise from many weaknesses in infrastructure design and operations, including the lack of generalization in our national critical infrastructure systems, which stems from insufficient security education and training programs. Ongoing security education and training must be a prerequisite for all personnel working on these national critical infrastructure systems. The complexity of infrastructure systems makes it difficult for organizations to effectively manage enterprise security. Non-uniform infrastructure system design and operations across the enterprise network can complicate security operations.

Cyber vulnerability includes the many security weaknesses that can lead to a cyber attack. Cyber vulnerability is comprised of Internet-based data communication and its associated susceptibilities. Cyber attacks exploit these vulnerabilities to compromise the target assets, such as the national critical infrastructure systems. There are many tools used in cyber attacks. Some of these tools are freely available on the Internet as downloadable software. These tools can cause devastating consequences to our society if used against our critical infrastructure systems. Examples of cyber attacks include the following:

- A distributed denial of service attack (DDoS) is when distributed systems participate in attacking a specific target synchronously or asynchronously.

- Worms replicate themselves without dependence on any agents or applications. The objective of worm attacks is to saturate the target's resources by replicating themselves throughout the networks and attacking network resources.
- Trojan horses are hidden behind seemingly normal software in order to exploit the target. Their intents, when executed, are malicious.
- Viruses replicate themselves by attacking, infecting, and modifying other computer software. The attacks could vary from removing the target contents to changing the integrity of the data.
- Backdoors can be left hidden and accessible to exploit the target behind the front-end software. The intention is only known to the perpetrator, who will take advantage of the backend channel for a later attack.

National critical infrastructure

'National critical infrastructure' broadly refers to a nation's physical and virtual, privately or publicly, interconnected systems that are essential to national security, economy, public health, and safety (Khalamayzer 2012). These infrastructure systems can be naturally existing or man-made, such as groundwater resources or electrical power grids. Many of these critical infrastructure systems are considered mission essential to the nation's security and the citizens' well-being. The US National Critical Infrastructure includes sectors such as banking and finance, the chemical industry, commercial facilities, communications, manufacturing, hydraulic dams, the defense industrial base, emergency services, energy, government facilities, healthcare and public health capacities, information technology, national monuments and icons, nuclear reactors, materials and waste, postal and shipping, transportation systems, and water systems (Amoroso 2013; NATO 2014).

If a cyber adversary attacked any of these critical infrastructures, the impacts on national security, the economy, and human lives could be devastating. Further, attacks on any of these infrastructures can have a ripple effect from one system to another, as they depend on each other's resources and protection. National critical infrastructure is tightly integrated with other networks, and should a failure occur on any part of the networks, the impacts would be far-reaching. For example, if a power grid goes down, the water systems could not operate, the health systems could not provide adequate services to patients, the transportation systems could not move shipments from one place to another, and agriculture systems could be crippled in delivering foods to our local grocery stores.

Whether these systems are physical or virtual, our computer systems provide crucial inputs to these critical systems. And protecting these computer systems is a paramount job for all of us (Keil 2014). Furthermore, private sectors and the government protect many of these critical infrastructure systems. The government has categorized these systems as national critical systems without which our society would suffer tremendously. As these national critical infrastructure systems are closely interconnected in cyberspace, we must protect the infrastructure for the nation to survive. Our duties and responsibilities as security professionals include identifying the national critical infrastructure systems and protecting them from any type of cyber attacks (Amoroso 2013).

Cyber attacks on critical infrastructure

Cyber attacks on any of these national critical infrastructure systems could cripple the nation's economy, financial markets, defense systems, healthcare industry, and others. We must protect these national critical infrastructure systems by detecting, preventing, and deterring any type of cyber attacks (Amoroso 2013). The Department of Homeland Security (DHS) assessed the national critical infrastructure systems for their security posture and risk factors against cyber attacks. Findings of these assessments reported that our critical infrastructure is vulnerable to cyber attacks, and we have much work to do to protect these infrastructure systems (Goldman 2012).

If a cyber attack targets the electric power grids, the resulting effects of the loss of electricity can be overwhelming to our nation's security, health, and economy. India recently experienced a power grid collapse in 2012, and the ensuing national crisis was a devastating economic chapter in India's history. The surrounding community that had relied on the power grid for their survival experienced horrific lifestyle consequences. The crippling power failures caused heavy traffic backed up at major highways as power outages left half of India without power. When India was attacked by its second large-scale power failure, it left over 600 million people without electricity or transportation networks.

This type of power grid failure could be caused by a cyber attack, in which case the consequences and impact could be much more devastating to the nation. When cyber vulnerabilities are exploited by cyber attacks, the attack payload could be targeted to specific organizations or users. They can also impact a large population covering a large territory. This type of attack is known as cyber terrorism, causing terrorism using the Internet. We must seriously consider how to detect, prevent, and deter against these cyber attacks (TNN 2012). Any discovered vulnerabilities on these critical systems must be mitigated before they can be compromised and cripple the nation's economic stability.

Cyber Attacks on Saudi Oil Refineries

The Saudi Arabian oil company Saudi Aramco, in August 2012, confirmed that it had suffered cyber attacks on its oil refinery sites, resulting in a loss of operational capabilities (Bronk 2013). Although Saudi officials could not confirm who had carried out the attack, the impact on the Saudi oil refinery has been devastating (Perlroth 2012). The attack was believed to have targeted the Saudi economy rather than a specific refinery site. Saudi Aramco confirmed that the cyber attack on its oil refinery aimed at the company's economic health rather than incapacitating one specific oil refinery site. Over 30,000 computer systems were affected by the attack. The perpetrator intended to cripple the refinery line to the mainland and interrupt the economy (France Presse 2012). This cyber attack demonstrated the impact of cyber attacks to the national critical oil supply line and the ability to seriously disrupt the economy.

Cyber adversaries

There are three categories of adversaries working against the national critical infrastructure systems: the internal, the external, and the supplier adversaries (Amoroso 2013). Of these, the internal adversary could be the most threatening, as they are already usually trusted insiders with access to critical infrastructure systems. These internal adversaries may possess administrative privileges to disrupt network resources using internal software, computers, and networks. The other two threat vectors are both external and must go through the perimeter protection and any other protection layers in place to attack national critical infrastructure systems.

External threats are any attacks initiated by external sources, such as those perpetrators with malicious intent to harm the target. These threats present various motivations, skills, and payloads. These external threats to national critical infrastructure can be perpetrated using remote access, system administration, or normal use. One of the external attack sources is the supplier adversary. Supplier attacks are carried out through the supply chain from the supplier of materials or components, extending through the manufacturing process to the distributor, retailer, and ultimately to the consumers. These cyber attacks aim for cyber vulnerabilities to exploit the target's data confidentiality, integrity, and availability (Sutton 2013).

Specific examples of cyber attacks

Malware attacks can be carried out by any of the above adversaries. Malware attacks are carried out in the form of viruses, worms, and Trojan horses with the intention to exploit the target's confidentiality, integrity, or availability. For example, malicious worm attacks do not need any application or the user's help to propagate. They can exploit their payload without anyone interacting with them (Johnson 2014). Worms self-replicate as long as they find network connections. Once infected, malware can damage the target computer boot sector, data files, and other software, thereby further damaging the computer operating system, file system, and critical databases. The target system could suffer damage to confidentiality, integrity, and availability (Dunham 2009). Using these malware capabilities, adversaries attack users as follows:

- E-mail messages masking the sender's identity attack the recipients of the e-mail messages.
- Malware can use instant messaging (IM) to send out attachments.
- Malware can be shared on a file server for users to click on them.

One of the most publicly discussed cyber attacks using malicious software is that of the Stuxnet worm. The Stuxnet worm has significantly changed the nation's cyber-security posture. Nation-state-sponsored cyber attacks were not as publicly discussed prior to the Stuxnet worm attacks. The Stuxnet worm exploited the target in the following manner (Amoroso 2013):

1. Infection – It infected a computer system via a thumb drive, e-mail, or a file share to access the internal network.
2. Search – Once it infected a computer system, it searched its target systems to further attack.
3. Update – It then updated its payload from the Internet.
4. Compromise – As it had been designed to do, it compromised the target system's vulnerabilities.
5. Control – As it took over control of the compromised system, it controlled the target system's information, specifically the operation of the centrifuges.
6. Deceive and destroy – While controlling the target system's logic, it eventually destroyed the target system.

The Stuxnet worm has brought attention to how industrial control systems can be used to cause physical damages to such facilities as power plants, dams, and other critical infrastructure. This tactic may allow an adversary to cause physical and economic damage to a targeted country without launching a military operation. The Stuxnet worm was detected around worldwide computer systems, infecting specific hardware and software systems to attack. Stuxnet was discovered affecting Iranian centrifuges in a nuclear enrichment reactor. With the help of North Korea, Iranians dissected the attack vector and discovered the CIA as the mastermind behind the virus. This attack was one of the first known cyber attacks (Kushner 2013).

The Flame worm was discovered after it had already been operating for some time. Iranian officials and researchers believe the Flame worm was designed by a nation-state, but they don't know who actually carried out the work. Many other sites and nations were infected by the Flame worm. More than 600 specific targets, from individuals and businesses to academia and government institutions, were hit by the malware. Syria, Lebanon, Saudi Arabia, and Egypt were also infected by the same malicious attacks (Hopkins 2011).

Attacks similar to Stuxnet and Flame have surfaced to target various European nations. A worm called Duqu was discovered infiltrating European companies to spy on the target systems. These attacks have shown how vulnerable our systems could be if unprotected, especially by these malware attacks. This new Duqu worm, like Stuxnet, is also a sophisticated worm with a very complicated attack payload. As with other malware, this worm replicates without depending on any other agents. Therefore, it can travel to any connected destination. Although the Duqu worm appears to have been collecting information rather than destroying the target infrastructure, its characteristics are similar to the Stuxnet worm and alarmed many people who speculated that the worm had been designed by the same people.

Cyber attacks on power utilities

Many of the current power utility systems are controlled by computer systems that are interconnected with the Internet. When controlled by the Internet-interfacing computer systems, certain cyber vulnerabilities could exist by virtue of the Internet connections. For example, certain utility power plants and electric grids are at risk of exploitation by cyber attacks. The DHS and the private sectors are doing everything they can to protect the country's utility infrastructure, but the current DHS risk assessment shows the protection is inadequate to prevent cyber attacks. We all fear power outages brought on by natural threats such as tornadoes, hurricanes, snowstorms, and the like. However, cyber attacks on power grids can wreak havoc on our systems of transportation, agriculture, health, water, and communications. We must do all we can to minimize harm by building safeguards in the form of redundant systems, fault-detection systems, proactive monitoring systems for operational status, and by performing routine maintenance (The Gawker Media Group 2014).

Cyber attacks on financial infrastructure

Cyber attacks on banks and financial institutions could cripple the economy. Interruption of our banking infrastructure could lead to a ripple effect impacting other industries, companies, and individuals (Folery 2012). We must prevent, detect, and mitigate any existing security vulnerabilities discovered in the financial industry. These security vulnerabilities exist due to the lack of defense-in-depth, insecure software developed without adequate safeguards embedded into the software development phases, or inadequate operations security. To minimize any security vulnerabilities, a proven risk management life cycle must be implemented to mitigate any risks arising in the financial sectors. Risk management is a life cycle of managing risks by asset inventory, assessing risks on these assets, mitigating identified risks, and monitoring these risks. Therefore, continuously monitoring the financial infrastructure must be implemented to minimize any impacts from these attacks (Khalamayzer 2012).

On 28 March 2013, the American Express website went offline for at least two hours during a distributed denial of service (DDoS) attack. A group calling itself the Cyber Fighters of Izz ad-Din al-Qassam claimed responsibility for the attack. American Express experienced intermittent slowness on their website that disrupted customers' ability to access their account information (Perez 2012). The American Express DDoS was part of a new wave of attacks started two weeks prior by the Izz ad-Din al-Qassam group, which launched a larger campaign targeting US financial institutions (Booton, Egan and Samson 2012). The group's alleged goal was to force the take-down of an offensive YouTube video or else extract an ongoing price from American banks as long as the video stayed up, which could be indefinite. These attacks are also part of a larger trend of disruptive and destructive attacks on financial institutions by apparently politically motivated groups. This type of politically motivated cyber attack could cripple a larger scale of cyber attacks using more Advanced Persistent Threat (APT) attacks (Rothman 2012).

An unidentified group of attackers outside of the country attacked JP Morgan Chase & Co., one of the biggest US financial institutions. The attacks targeted one of the most vulnerable servers in such an organization: the web server. The web server has traditionally been one of the most targeted servers in an organization due to its open ports designed as such so that client applications can establish a socket connection. Having established a socket connection, further exploitations by malicious attackers could be possible (Canham 2012). In 2012, attackers flooded bank websites with malicious traffic, rendering the target web pages unavailable to consumers and disrupting transactions for hours at a time. Such denial of service (DoS) attacks are commonly used against web servers to overload the processing resources on the target web servers, rendering them inaccessible to customers (Gilbert 2013).

The financial market heavily utilizes the cyber infrastructure to assess market fluctuations and responses. The market supports many clients and customers by providing accurate and near real-time data. Financial decisions are made based on data analysis from the market, and many transactions are time-sensitive. A fully operational infrastructure is critical to providing accurate and meaningful data. At any given time, if the financial infrastructure experiences an unexpected outage, the resulting damage to the financial market could be devastating. As security professionals, we must be able to assess risks to the US financial market infrastructure for any potential interruptions by cyber attacks (Kumar 2012).

Cyber attacks on defense systems

Cyber attacks on national defense infrastructure could cause devastating effects to the entire nation. The national defense infrastructure not only operates to effectively protect the national defense systems, but also collaborates with other private-sector entities to achieve common security goals and objectives. Many of the defense systems are maintained and operated by the private sector, so governmental collaboration with the private sector is mission critical. The national defense infrastructure includes systems to protect the nation's air space, water, land, and cyberspace. The Air Force, Army, Navy, Coast Guard, Marines, the National Guard, and other defense branches rely on electronic and physical defense systems. The national defense infrastructure includes nuclear weapon systems, cyber warfare capabilities, various traditional weapons systems, and so on. Any interruptions to these systems can cripple national security in one or more specific areas. If attacked, the nation's defense systems could be debilitated and provoke devastating cyber conflicts between nations.

The US defense systems are protected by government agencies and private sectors working together toward common security goals and objectives. However, many of the defense infrastructure systems could improve their security positions by assessing security risks, mitigating security vulnerabilities, and continuously monitoring security postures. In doing so, security professionals would be able to assess and recognize cyber vulnerabilities in order to mitigate them. Security vulnerabilities may stem from older, legacy defense systems that must be replaced or modernized. Legacy systems were not designed with security embedded in their system development life cycle (SDLC). Furthermore, some of these systems have deteriorated over the past many years of use, and replacing them with more modernized systems is an important national security strategy.

Cyber warfare

An analysis of the distribution of cyber attacks on the US government shows there is a consistent pattern of cyber attacks each month. Cyber crime, hacktivism, cyber warfare, cyber terrorism, and cyber espionage are considered cyber attacks (Amoroso 2013). An attack must succeed only once in order to possibly gain access to our critical data, while our defensive measures must work 100 percent of the time to deter adversaries. The *Washington Times* reported that the US weapons systems had been compromised by Chinese cyber attacks (Arguilla 2011). The attacks accessed classified data and compromised data confidentiality. The breached defense system is a national security concern as it might have released secrets on US weapons systems to our adversaries. When critical information and technologies are released to foreign nations, it endangers national security as the adversary could use the information against us. As these data could be critical to national security, we must safeguard them at all times from adversaries (Clayton 2013).

A DoD report indicates that there have been a growing number of black-market sales for zero-day vulnerabilities that can be destructive to our industrial control systems (Magnuson 2013). Zero-day vulnerabilities are discovered as new vulnerabilities, and there have been no patches available as of yet. If these vulnerabilities were discovered by adversaries, they could compromise these zero-days or sell them to others for profits. An example of zero-day attacks was the Stuxnet worm that attacked the Iranian SCADA systems attached to Iran's nuclear program networks. The Stuxnet worm disrupted the normal operation of centrifuges used to enrich uranium (Magnuson 2013).

The cost of cyber attacks

In response to cyber attacks, we must leverage systematic emergency responders who are trained well in dealing with these types of cyber attacks. The role of an emergency manager is to assess the emergency situation and respond accordingly. The emergency manager determines the best course of countermeasures and safeguards to deter cyber attacks and works closely with the security team to coordinate counterterrorism activities or cyber attacks. Cyber war is a very delicate topic to discuss if it is indeed an act of war. There has not been a concrete national policy or standard to define what constitutes a cyber war. Lawmakers have not come up with this new cyber law initiative to support the decision on cyber war. This is rather a new initiative and a clear cyber law or policy must be passed in order to properly respond to these cyber attacks.

The cost of cyber crime is steadily increasing at an alarming rate. A study shows that cyber crime has a real impact to our economy, costing $8.9 million annually (Murry 2012). Cyber crimes include identity theft, money laundering, and other cyber-fraud activities. Computer criminals are becoming more computer savvy and entice non-technical computer users to participate in illegal cyber activities.

National critical infrastructure protection plan

As the critical infrastructure systems are controlled by computer systems, protecting computer networks from cyber attacks is mission critical for national security. Industrial control systems (ICS), like the Supervisory Control and Data Acquisition (SCADA) systems, manage the nation's critical infrastructure. As these industrial control systems are managed by computer systems, computer vulnerabilities do exist. Any interruptions to these SCADA systems could mean interrupted water systems, power grids, transportation, and agriculture systems. Water control systems are interconnected with the Internet; therefore, they are vulnerable to cyber attacks. When these control systems are interconnected with the Internet, they can be remotely managed to collect industrial control data. A recent cyber attack on an Illinois water utility revealed that a Russian-based Internet Protocol (IP) address initiated a cyber attack on the system after stealing a credential from a company that serviced SCADA systems. This cyber attack incident prevented the system from delivering clean water (Kavhaz Center 2013).

In 2012, as many as 198 cyber attacks against industrial control system/SCADA systems were discovered (Hale 2013). The security research discovered that many of the ICS in the US are directly interconnected with the Internet, exposing vulnerabilities to perpetrators. As such, these systems are easily searched, targeted, and attacked. Information on various ICS can be easily found with Internet search tools, targeted, and then compromised (Hienz 2011).

Another example of what cyber attacks mean to these critical systems was illustrated by an incident that affected the Iranian oil industry in recent years. Iran shut down Internet access to its oil terminals following a cyber attack with a virus called 'Wiper.' The virus wiped many of the hard drives in the Iranian oil industry computer systems, and the oil production data was destroyed.

Cyber attack search for vulnerabilities

There are four distinct attack stages adversaries go through to successfully carry out their exploitations. First, scanning discovers the target network topologies, open ports, running services, and security vulnerabilities. Adversaries can use many types of scanning modes to determine the target networks, servers, and routes to the target. Second, based on the scanning results and the network discovery, an attack on the target can be planned. Third, once a target has been identified and scanned, the actual exploitation can be executed to complete the attack. When the target has been exploited, the network assets are exposed to the attacker. The perpetrator exploits the vulnerabilities discovered from scanning the network.

Protecting against future cyber vulnerability

Cyber vulnerabilities are becoming increasingly prevalent as more computing devices are available on networks and we process more business transactions online. While accessing our proprietary information across the network, it is important that we do not expose sensitive information to adversaries. In particular, national critical infrastructure systems are most critical as the impact of their loss in confidentiality, integrity, and availability could be devastating to the national security, economy, and even human survival. Finding cyber vulnerabilities and mitigating them early is critical to our success in cyber protection.

The potential bright spot for those seeking to stop the proliferation of these kinds of attacks is that there are only a small number of experts who are capable of finding vulnerabilities in SCADA systems, and fewer still willing to exploit this knowledge. These experts must have a high level of technical skill to find these SCADA-specific vulnerabilities, and not many possess the capabilities to compromise these critical systems.

However, there will always be persons or entities seeking to do harm, and a three-pronged strategy of mitigating such attacks has been suggested. The first prong is to strengthen information sharing among the vulnerable critical infrastructure sectors. Second, when security vulnerability is discovered, one should share it with others to minimize any further impacts on this newly discovered vulnerability as any unpatched vulnerable systems could potentially be compromised. Third, the government and private sectors must work together to enhance the overall security posture by sharing security information, promoting education, and training.

Unfortunately, many of the technicians who operate SCADA systems may not be cyber-security experts; therefore, an ongoing cyber-security education and training must be in place in every organization (Gallagher 2013). There also needs to be stronger international cooperation between law enforcement agencies to catch those who are involved in the black market of selling zero-day attacks, which could be sold to foreign nations for a high profit. There are countries that lack strong legal systems to prevent criminals from engaging in these types of activities. The first step is to make sure that such laws are in place, and then to enforce these laws with severe penalties (Magnuson 2013).

Conclusion

Cyber vulnerabilities can expose national critical infrastructure systems to possible cyber attacks. Therefore, it is imperative to remove all discovered cyber vulnerabilities to reduce any potential exploitation. In order to mitigate cyber vulnerabilities, we must be able to identify where cyber vulnerabilities may exist in our cyberspace. Continuous monitoring and assessing our assets and security controls can reveal the cyber vulnerabilities for the national critical infrastructure systems and other mission-essential systems. Strategically, security must be well balanced between both operational and audit practices. Constant audits to check current security settings and effectiveness are a part of the necessary security measures we need to protect cyberspace.

12
Opportunities and Security Challenges of Big Data

Zal Azmi

Introduction

The use of *Big Data* to analyze occurrences, issues, phenomena, and the like in a variety of fields and disciplines is a rapidly developing trend in our increasingly interconnected world. References to Big Data are now a normal part of the daily discussions and technology discourse of average consumers as well as foremost experts in a range of industries. In an environment where all devices are interconnected and information is exchanged without the need for an intermediary, the rapid growth of Big Data is unavoidable and the scale of growth is unpredictable. Over the last decade, Big Data has demonstrated a remarkable capacity and utility to help different business communities make informed decisions, save money, and improve policymaking. The use of Big Data throughout the federal government is well documented, with several initiatives that utilize data to inform procedures, systems, and situational awareness in the work of a variety of federal agencies.[1]

As of 2012, any given US federal agency had an average of 1.61 petabytes of data (1024 gigabytes 5 1 terabyte and 1024 terabytes 5 1 petabyte) stored, and that number is expanding rapidly.[2] A medium-altitude unmanned aerial vehicle (UAV) like the MQ-9 Reaper can collect more than 10 terabytes of data, which equates to the complete printed collection of US Library of Congress, in a single sortie. While the evolution of Big Data has introduced new computing paradigms, tools, and technologies, and new thinking about what data can do, legitimate concerns remain about the nature of information security and the usage of Big Data to solve immediate and long-term problems. In this chapter we will explore the importance of Big Data in solving complex and emerging problems that rely on combining a massive amount of data from different data sources in real time while maintaining the security of the data sets.

Definition

There is no single definition for the term *Big Data*. Big Data is not just a technical description; it encompasses analysis and issues related to privacy, economics, politics, security, and academia. For the purpose of this chapter, Big Data can be conceptualized as any data whose size exceeds the cost and capabilities of the current technologies and requires new methods and technical tools and solutions to make sense of this massive amount of information. This definition is inclusive of technologies such as NoSQL, MapReduce, and machine learning tools. Big Data is often thought of as possessing the *four Vs*: volume, variety, velocity, and value. Also, there are additional dimensions to Big Data related to the tools and technologies required to collect, normalize, correlate, analyze, and process it.

Volume (size)

As technologies have evolved in sophistication and capability, the size of what constitutes Big Data has also evolved. During the period of the creation of statistical methods, the largest data set amassed and analyzed could not have been more than 50 data points. Several factors contributed to the lack of additional data points, such as limited connectivity, the lack of information in digital format, and the lack of hardware and software to collect and combine the data points. Today the Acxiom Corporation maintains information on 500 million active consumers globally and collects over 1,500 data points per consumer (The New York Times 2012). At the dawn of personal computing, Big Data was considered more data than could fit onto a hard drive. In the 1980s, the IBM 3380 mass storage system could hold at least 100 gigabytes of data and it was priced between $97,000 and $142,000.[3] Today, a modern personal computer with a one-terabyte hard drive and storage device can be acquired for less than $100.[4] Cloud service providers are now selling a gigabyte of storage on publicly available clouds for less than seven cents. The ability to obtain cheap, efficient, stable, and always available storage has led to a proliferation of data gathering to satisfy different business needs and outcomes in both the public and private sectors.

For example, in 2012, Facebook ingested 500 terabytes of data daily to address the needs of its data operations.[5] To maintain and increase their client base, Facebook relies on positive user experiences, which necessitates providing real-time and relevant information to their customers, such as making all relevant information available about individuals before a user befriends them. By collecting survey responses from clients, Intel links Big Data capacity to organizations that generate 'a median of 300 terabytes of data weekly.'[6] The federal government, and specifically the intelligence community, can generate much more than this through their vast networks and tools of data collection and storage.

Variety and complexity

Big Data is frequently unstructured and contains a mix of different formats. It could come in the form of audio, video, images, and text. The complexity becomes more pronounced as collected data is normalized to reduce duplication and redundancy in order to efficiently and quickly analyze the data sets. A single surveillance instance of an individual (for example) could produce a complex data set that includes geolocation data, telephony data, text messages, voice recordings, pictures, and videos. Additionally, the Internet of Things (IoT) and the type of data that the sensors generate are unique and different. IoT includes every smart device with an Internet Protocol (IP) address that uniquely identifies it on the Internet and is inclusive of smartphones, home appliances, monitoring systems, and so on. Especially on such a large scale, these data sets are difficult to process with traditional tools in a way that preserves the permutations of linkages between them.

Velocity

The explosion of social media tools, along with the capabilities and wide reach of IoT, has given incredible speed to the generation of data. According to Gartner (2012), in a single minute we send 204 million e-mails; stream 61,000 hours of music on Pandora; perform 20 million photo views and three million uploads on Flickr; type 100,000 tweets; conduct six million views and 277,000 Facebook logins; and execute more than two million Google searches. Cisco forecasts that there may be 50 billion devices that could potentially be connected to the Internet by 2020.[7] By 2030, the quantity of sensors is expected to reach one trillion, which would make data generation by the IoT the dominant part of any Big Data system.[8] With the low cost of storage devices, this data will be collected, stored, and analyzed to address existing and future business needs.

Value

The value of the data refers to the provenance of the data. Making appropriate decisions based on data sets requires traceability to the original sources. Since Big Data references and combines data from different sources, the quality may be good, bad, or unknown depending on the sources and the state of data. In many cases, the value of the data may be determined by its completeness, definition or lack thereof, timeliness, relevance, and so on.[9]

Tools and technologies

The collecting of data is much easier than the process of analyzing the collected data and making sense of it. As Jacobs (2009) argues, it is simpler to get data into a database than out of it.[10] Data becomes Big Data when the current technological capabilities do not scale to deal with the size of the data set. It is imperative to use a combination of tools to process (collect, normalize, correlate, analyze, and visualize) the data. A single tool is unlikely to be sufficient for Big Data analytics.

Data sets have inherent temporal and spatial dimensions, which can greatly increase the difficulty of an analysis. Algorithms that function efficiently with small data are often interminably slow, and their platforms are inefficient when dealing with large data sets. Software applications frequently have a hard limit on the size of data that can be handled. The same is true for the hardware platforms. Memory, CPU, or other limitations could also affect the software applications. Enhancements in cloud-based processing and open-source software such as Hadoop will counter some of the basic limitations of the most widely used analysis tools. However, additional enhancements in capacity and continuously evolving technology are needed.

Big Data initiatives will expand in scope and complexity as huge volumes of data are generated and consumed by individuals, IoT, and social networks. Much of the data flowing from these sources will be unstructured and transient in nature. It is important to note that the transient nature of the data will demand more real-time processing if the data is to prove valuable. In general, data is considered as a perishable commodity and loses its value over time. At times of emergency or when there is an immediate threat to public safety, the value of data initially increases rapidly, and then decreases exponentially. However, after the immediate value of the data is utilized or exploited, the same data may be published later as an open data set and reused for public or commercial activities that create new services and insights from identified trends. Extracting the maximum value from the relentlessly increasing volume of structured and unstructured data acquired from internal and external sources continues to stretch the traditional information management practices of many organizations.

Similarly, Big Data's impact on enterprise architecture addresses the fact that Big Data has the power to adversely impact an organization's operation by providing frequent or constant disruption to existing business models. The use of enterprise architecture can promote the resilience and flexibility needed to adapt to rapid changes and manage ongoing risks. The advent of technologies that can deliver significant but short-lived return on investment has placed pressure on traditional procurement policies and practices as well as project management and will demand that agile procurement operations become the norm for government agencies.

Big Data and national security

The Internet of Things has fundamentally transformed the way humans and sensors interact with each other and generate data. It has increased the volume, variety, and velocity of data that may be potentially relevant to a myriad of national security concerns. The concerns include counterterrorism, espionage, criminal investigation, weapons of mass destruction proliferation, and information network security.[11] Many of these problems pull together heterogeneous and unstructured data from multiple sources and do so at a near real-time pace.

A case in point may be the example of tracking a domestic terrorist within the United States. This individual, like many others, has a job, uses communication devices to keep in touch with family members or accomplices, maintains a home address, conducts financial and banking activities, and has friends and relatives. Each one of these facets of life provide an opportunity for the law enforcement community to collect information about the individual in real time and analyze it for threats and prevention of attacks against the United States.

For instance, intelligence analysts and information security professionals deal with Big Data on a daily basis in order to prevent or counter various threats to business, financial, and governmental operations. The impact of failing to exploit defense-related data in particular could be disastrous, with governmental consequences more far reaching than those typically faced by the private sector. Members of both the intelligence and law enforcement communities crawl through masses of data looking for unique threat signatures that change constantly and rapidly, forcing rapid analysis and updates.[12] Couch and Robins (2013) argue that military commanders' expectations of real-time information to inform their decisions are growing. They continue that in the field of cyber security, Big Data is being used to spot advanced persistent threats

(APT) against government information. The security tools deployed to protect the information networks of government agencies produce massive amounts of information about the activities of their users and the defenses of the devices, all of which must be collected and analyzed. The electronic signature that individuals leave behind is massive and could include all of their online activities (e-mail accounts, Web surfing, downloading or uploading malware, visiting elicit websites, banking, travel arrangements, tweets, Facebook, Flickr, and so on).

Big Data and privacy

In the context of national security, there are multiple intersections between Big Data and privacy. Big Data can exacerbate the existing tension between civil liberty advocates and the national security organizations because Big Data capabilities allow a large amount of personal data to be collected about individuals, stored for long periods of time, and analyzed with tools and technologies to discover new information. The legal environment in the United States mandates strict control of the protection of personally identifiable information (PII) and information about individuals who are not under investigation as a national security threat. The increased collection power has given rise to a very untenable situation where those who have access to the data may not be properly overseen.

An example of this lack of oversight is the disclosure that a National Security Agency (NSA) employee used the Agency's data-gathering powers to collect information about and spy on ex-girlfriends.[13] Other examples involve the disclosures of national security information to the world by both Edward Snowden and Private Manning. These incidents have made it clear that individuals on a network can potentially have access to a large amount of information either directly or through manipulation of access privileges. Human beings still must safeguard Big Data, and our security model is only as strong as its weakest point. Those who work with all types of data must be trained in the appropriate use of the capabilities and be monitored for data misuse and abuse.

Internal and external to the government, Big Data tools and technologies are used to collect, store, and analyze various aspects of US national security policy. If this data was lost or compromised, potentially millions of people associated with national security in the US or abroad may be affected. Some security concerns can be mitigated by striking a balance between security and privacy interests through implementation of special tools and policies. Financial institutions have managed to find effective strategies for protecting data privacy while maintaining usability in multilevel security approaches. The reduction in the number of people with systems administrator privileges, separation of duties, multilevel authentications, encryption of data at rest and in motion, and additional monitoring of administrator privileges are but a few of these strategies.

Big Data and information sharing

The government, private sector, and national and international organizations constantly need to share mission-related and/or business-relevant information among national and local assets. In many cases, informed communication is necessary to prevent or neutralize an attack through real-time collaboration and information sharing. This communication and sharing of Big Data, along with the tools and techniques required to analyze it, require human resources talent that is difficult to find. By 2018, a McKinsey Global Institute report argues that the United States might face 50 to 60 percent talent gap between the supply of people with experience with Big Data methods and the demand for analytic talent and data managers.[14] It is also important that, in order to effectively share data, national assets have the personnel to develop secure information systems that are also usable. However, it is not just lack of skilled data managers and analysts that can hinder efficient sharing of information.

In addition to developing and recruiting talent, organizations may have to also address technical challenges, such as problems with parallel processing, latency, lack of atomicity, consistency, isolation, and durability.[15] Most of these issues result from a general lack of financial resources, and most organizations still own and operate ten-year-old systems. The future of data collection will require upgrades of current systems, sensors, and hardware and software architecture. Enhanced analytic servers and high-performance computing servers and applications will be needed. Individual organizations may not have the resources, talent, or budget to deal with the challenges on their own. Coordination must be encouraged and facilitated, if only to take advantage of the existing available talent and increasingly austere budget environments.[16] Big Data may compel increased information sharing and collaboration as a cost-saving measure.

The lack of common information systems between partners also makes it difficult to share critical information, especially in a combat environment or crisis situation. For example, organizations both internal and external to the government are frequently tasked with finding intelligence related to the same mission, but there is no unified data source or network to provide intelligence support as of yet.[17] Many believe that Big Data could act as the catalyst for increased interagency coordination and information sharing. No single agency will have the capabilities and, perhaps more importantly, the funding to provide full operational capability by itself. The sheer volume, velocity, and variety of the data, coupled with lack of the most cutting-edge tools and technologies, will present continued challenges to agencies. While experienced data users such as the NSA may have the technical expertise and budget to handle these challenges, state, local, tribal, and other smaller government organizations may not. Information sharing and cooperation may provide a way for less-resourced organizations to benefit from Big Data without needing to invest in redundant capabilities.

One example of successful information sharing is being implemented by the Federal Bureau of Investigation (FBI). The FBI's National Data Exchange (N-DEx) is an example of purposeful Big Data information sharing.[18] The FBI N-DEx program was launched in late 2000 as a mechanism to share law enforcement and criminal justice information across multiple jurisdictional boundaries. It has more than ten terabytes of data and serves a range of law enforcement agencies around the country. Each of these agencies (as well as Interpol) can upload data and use the database for law enforcement investigations; pretrial release investigations; and investigations associated with the processing, presentencing, and correctional institutionalizing of suspects.[19] The data in N-DEx can be searched via the use of keyword queries or through a search engine to perform correlation and resolutions analytics. The N-DEx application has enabled many smaller law enforcement organizations to benefit from the FBI's investment in Big Data.[20]

Big Data expands the types of data that can be collected, and information sharing must follow both legal and operational regulations to ensure that data privacy is protected. There are numerous different legal regulations and more are being authored, their applicability depending on the types of data, classification level (need to know vs. need to share), and authorities involved (for instance, walls of separation between domestic surveillance and foreign surveillance). In battle domains, Big Data can only be gathered from a defined battle space to be effective.

From an operational perspective, not all data collected is useful. Foreign and domestic intelligence organizations gather data from many sources including, but not limited to, chatrooms; message boards; e-mail accounts; phone calls; social media outlets; and network assets such as servers, switches, and routers. While this constitutes large amounts of data (Big Data), most of it is innocent and unrelated to the problem being examined. Sharing of information that is not related to defined security goals and objectives and that does not comply with legal regulations violates the privacy of the data owner.

For the United States, any information-sharing strategy related to national security and critical infrastructure protection must also take into consideration that more than 80 percent of the US critical infrastructure is owned and operated by private-sector corporations. This formal and informal public-private arrangement necessitates a sharing between all players, private or public. Yet, the governmental intelligence community places a low priority on sharing information related to threats with the private owners and operators of the critical infrastructure.[21] Part of the issue is related to the sensitivity of the government data because very few private-sector entities have the facilities or clearances to deal with classified information. The private sector is also reluctant to share because leaders want to protect their stock prices and reputations of their organizations. There will not be many organizations in business if the word gets out that they cannot protect consumers' information or defend their networks against cyber attacks. Together, the government and private sector must reach a delicate balance in sharing information about threats while protecting national security, consumers, and organizational information.

The fact that massive amounts of data are often stored and processed in a single location makes Big Data storage an attractive target for cyber (and physical) attacks. For any organization, an attack on its data infrastructure can be costly. For government organizations, the repercussions of an attack, whether internal or external, can be magnified and pervasive. For example, Google claimed that the Chinese cyber attack on their servers in 2010 targeted human rights activists' information (Washington Post 2010). According to ZDNet (2013), the actual aim of the hack was to gain access to a database which contained years' worth of information related to US surveillance and law enforcement activities. As more and more organizations increasingly move to the cloud for storage, it is imperative that security professionals develop adequate threat models and mitigation strategies to deal with the likelihood of a cloud attack.

The characteristics of Big Data require a massive information technology environment capable of handing the volume, the speed, and the types of data that need to be analyzed. This means large and fast computers with a lot of storage residing in a single location or interconnected information technology capabilities (clusters) across multiple geographically disbursed locations. The computational platform required by Big Data frequently means that organizations will have to deploy some type of distributed programming framework for efficiency. The MapReduce framework (Bhatotia et al. 2011), of which Hadoop is the most popular open-source framework, splits an input file into multiple batches and distributes them computationally across multiple clusters. A mapper reads the data, performs some type of computation, and then outputs a list of key value pairs. After this step, a reducer combines the values belonging to each distinct key and outputs the value. Using this framework greatly increases the speed of computation. The distributed nature of the system also enables an organization to build powerful computing platforms more cheaply by adding and taking away computing clusters as necessary.

Distributed programming frameworks are vulnerable to two different attack vectors: insecure mappers and unstructured mappers. An insecure mapper may exhibit data leakage and accidently reveal private information or return incorrect data. In a large cluster, especially the kind necessary to analyze massive quantities of data, it will be very difficult to identify which mapper is insecure. Unstructured mappers can be altered and their security could be compromised by malicious users or trusted insiders to snoop on a request, return false data, or modify scripts entirely. This may occur due to a change in the configuration of worker nodes, intercepting communications between master program and worker nodes, through the introduction of nodes to the framework that pretend to be part of the cluster, or a combination of these attacks. The introduction of rogue nodes is particularly difficult to detect in a cloud environment.[22]

Malicious mappers may cause significant damage to a program or system by altering the output. For example, an attacker could implant a malicious script in a MapReduce framework that is used to analyze the performance of jet engines for new military aircraft, compromising the results and potentially endangering lives. Rogue nodes can alter the nature of computations or steal data and store it in a location controlled by an attacker. Information sharing between law enforcement agencies could potentially be compromised in such a way as to leak information about persons under surveillance, including location data.

Big Data models are also confronted by problems related to cryptography. Secure communication and computation has been a major area of research for the past three decades, but Big Data provides an important research stimulus because conventional approaches to cryptography of these Big Data are inherently insufficient (Malkin 2013). Cryptography models were not built for the massive scale, complexity, and technological scope of Big Data. Distributed systems exacerbate these issues because in most cases the data must be encrypted at rest (while it is stored) and during transmission between devices. This requires a lot of processing capabilities to encrypt and decrypt the information in real time. Client processing must be run separated from each other, making access control more difficult (Rass and Slamanig 2013). This is required for better performance and faster access to information. However, the implementation of fine-grained user-roles–based access controls is still challenging in a cloud-computing environment. Clouds are still relatively new environments, and new and stronger methods of encryptions are needed to protect user information.

It is understood that despite the implementation of all security tools, standards, and best practices, certain attacks on Big Data will still be successful given the adversary's motivation and access to resources. Due to this eventuality, it is imperative that an ongoing and continuous audit of Big Data platforms be conducted. This audit should look at the changes to access controls, usage, transfers, data (updates and modifications to the original data), and the like. Personnel that have access to all necessary log information should complete audits. Adequate information protection and encryption should be in place to safeguard against data compromises. Only authorized and knowledgeable staff should have access to the audit process and results, and they should be authorized to view only the data necessary to perform their duties.

Data provenance presents another challenge in the Big Data environment (Glavic 2014). Provenance information explains the creation process and the origin of data through recording what transformations happened (from creation to present) and from which data items a piece of data originates. Data provenance has applications in debugging, trust models, and security. Without provenance information, it may be impossible for a user to know where the data came from and what transformations may have occurred, thereby undermining the value of the data.

In a Big Data environment, provenance metadata (information about data) will grow rapidly in complexity. For example, when a picture is taken by an iPhone, the additional information (metadata) that is collected could include the GPS location of the picture, the time and date that the picture was taken, the photographer's personal information, information about the device that took the picture, and other such details, all tagged to the photo. This information is usually not presented to the user, and many users are unaware that this information is actually stored on their iPhone. As Big Data is typically heterogeneous, it becomes difficult to define a common structure for the provenance of a data set. Big Data systems and platforms also may be computed in an ad-hoc fashion from different locations, and data items may be produced by transformation from different analysis. For security and efficiency purposes, it is helpful to put in place electronic systems that allow data to be monitored at a fine-grain level in ways that scale efficiency with the size of data. If done correctly, data owners can identify bottlenecks in the performance, scalability, and robustness of the data collection system.

As cloud computing becomes ubiquitous, more administrative, financial, business, and mission-related information likely would be placed in the cloud. The cloud accommodates the presence of multiple Big Data instances that could be targets of attack. An attacker who has gained control of a cloud may use the available computing power of the cloud to manipulate the instances of Big Data, spread malware, conduct denial of service attacks, or perpetrate spam. Data aggregations and access to multiple instances of Big Data for data mining and analytics purposes may also present a challenge, where the data sets by themselves are not sensitive, but the aggregation process and analytics results could expose sensitive information in a way that potentially violates privacy and security laws. Still, Big Data systems and corresponding analytics are often more secured than previous generations of data collection systems.

Big Data analytics can be divided into two groups: batch processing of data at rest and stream processing of data in motion. Data analytics is being leveraged for information security and situational awareness purposes because of the amount of data captured by automated sensors and system logs. The Cloud Security Alliance, a cloud security membership group, outlines the evolution of Big Data security analytics.[23] In the first generation, security architects realized the need for layered security. In the second generation, security information and event management (SIEM) aggregated and filtered alarms from many sources. The next generation will use Big Data analytics, thereby reducing the time for correlating, consolidating, and contextualizing various pieces of information. These continuous enhancements in technology ensure that Big Data systems can be more reliable and available than they are today, providing guarantees that queries will be processed to completion, indicating an additional advantage over traditional security systems.[24]

Big Data can improve security of other systems by providing real-time security monitoring. This introduces two challenges. First, how should an organization monitor Big Data itself? Second, how can this same infrastructure be used for data analytics? For national security assets, the answers to both questions are critical. As mentioned above, real-time security is a challenge because the volume of data very likely generates a number of security alerts. Most of these alerts will be false positive, and a well-tuned monitoring system will dismiss most of them. Human analysts will ignore the false positives, and the remaining alerts will in some ways scale with the size and velocity of data streams. A protocol that incorporates best practices can provide the necessary security monitoring and analysis.

Best security practices for unstructured data

As the amount of data increases, most relational databases will exhibit performance issues. According to EMC^2 (2012), when the size of data increases, relational databases suffer because they are not well suited for storing unstructured data, and it is difficult to implement certain basic queries using SQL and relational databases (for example, what is the shortest path between two points). This has prompted many organizations to move toward NoSQL databases. NoSQL provides methods and a wide variety of different database technologies that were developed to address relational databases' shortcomings in handling the rise in the volume of data, the different types of data, and the frequency in which this data is accessed, and performance needed.[25]

Currently there are more than 150 NoSQL database products available for use by organizations. NoSQL databases are more scalable and have superior performance over traditional relational databases. They are also more flexible because they do not demand that a schema be created prior to data being added. As the amount of data increases, 'NoSQL databases usually support auto-sharding, meaning that they natively and automatically spread data across an arbitrary number of servers, without requiring the application to even be aware of the composition of the server pool.'[26] Consequently, data and query loads are balanced across cluster of servers, so if a single server fails, it can be replaced without disrupting the applications. This feature is most useful in a cloud-computing environment, which requires nearly unlimited capacity on demand.

While NoSQL is better suited for Big Data storage, the database architecture itself is insecure. NoSQL's biggest advantage is also the biggest security risk: the dynamic data model.[27] Dynamic data modeling refers to a data store's (data warehouse) ability to adapt to the new changes in the data environment, such as the data usage pattern, data upload pattern, relationships among data sets, and other such characteristics. NoSQL has no inherent security for use by the data and designers. Developers must embed security in the form of middleware to protect the data stored in NoSQL databases. Additionally, NoSQL databases have weak or no encryption support for data files, weak native clients and server authentication, simple authorization, and are still vulnerable to JavaScript or JSON injections and denial of service attacks.[28] The security of files and data is not a feature of the NoSQL database itself, and there is still no well-defined security standard for NoSQL databases.

In contrast, traditional databases are developed with security in mind and have well-formed schemas that require adherence. Checks and validation protocols are built into the system to ensure proper operations. NoSQL is instead a dynamic database, and building the checks and balances must be anticipatory, because new attributes can be added at any point in the process. One challenge with this dynamic model is that NoSQL databases are still relatively new, and many users don't know how to configure them properly. Ming Chow, a computer scientist at Tufts University, notes that NoSQL instances place an emphasis on trusted environments.[29] It is a well-known fact that most users are not well equipped to cope with constantly changing IT platforms. Administrators like to implement the same queries and see very few changes in the database configuration. Unfortunately, NoSQL databases frequently differ on basic configuration, forcing database administrators to set up individual configurations. This may make information sharing between NoSQL databases more difficult when organizations use different databases, presenting a challenge for the computer scientists and compelling them to learn the new configuration and set up new queries to retrieve information.

Despite all of these challenges, the Big Data infrastructures and technical assets must ultimately be made secure by the personnel entrusted to use them. Information operators must have the appropriate skill sets, security credentials, administrative access level, and need-to-know clearance before conducting analysis of the security events or uploading sensitive information onto the infrastructure.

The future of Big Data

The future of Big Data depends on continuous technical enhancements and improvements. The National Research Council highlights three open problems for Big Data in the context of national security.[30] First, to make data smarter and more usable, streaming algorithms that can process data in one pass with low memory usage must be built. Second, tools for unstructured systems need to be built that are both secure and usable. Tools that have simple and easy-to-use interfaces will be particularly helpful, even assuming technical competency. Third, to follow along with better algorithms and better database systems, visualization tools are needed to allow analysts to quickly and efficiently understand data sets. Even in the new world of increased data literacy, analysts will be pressed for time.

The IoT will have the largest impact on the future of Big Data because it has three features that are embedded within the existing definition of Big Data. Most obviously, there are many sensors within the IoT, generating masses of data. The volume of data generated by these sensors cannot be predicted at the moment. Second, because sensors are different from one another, the majority of this data is unstructured. A cellphone that generates voice data and text data (unstructured data) will produce data that does not have a strictly defined schema, unlike a global positioning system (structured data) that does have a defined schema. Finally, the data from IoT is raw and not always immediately useful; it must be normalized and analyzed to extract useful and actionable information. In the future, new standards and technologies are needed to address the security and privacy of the data traveling through the IoT and stored in the Big Data systems and platforms.

Conclusion

Big Data has fundamentally altered the way we look at data collection and how we draw conclusions from such data. It holds great potential for unlocking new linkages in the sharing of information and real-time collaboration between national assets. The FBI's Next Generation Identification (NGI) will revolutionize the identification of criminals through use of the next generation of biometrics (fingerprints, facial recognition, palm prints, and iris scans). This information can be easily used with national and international partners to positively identify criminals.[31]

Big Data plays a major role in many fields of research, such as healthcare (specifically genome sequencing), cyber security, and intelligence (especially the open sources intelligence), among others. Unfortunately, Big Data has also created new challenges and risks to the data in terms of confidentiality, integrity, and availability. Big Data provides the opportunity for new attack vectors because of its use of distributed processing, presence in the cloud computing environment, and its massive scale. Additionally, the expansion of IoT and ubiquity of sensor data will present more challenges to security and data privacy.

As Big Data expands, the sources of data that are combined must be vetted for trustworthiness. Regulatory incentives and technical capabilities must be developed to ensure national security and business interests can be defended without compromising the privacy of the collected data. The security of Big Data platforms is imperative, and a continuous, disciplined monitoring routine must be in place to proactively identify and remediate potential weaknesses in data and system securities.

Notes

1 http://www.whitehouse.gov/sites/default/files/microsites/ostp/big_data_fact_sheet_final.pdf
2 http://www.meritalk.com/pdfs/bdx/bdx--whitepaper--090413.pdf
3 http://www-03.ibm.com/ibm/history/exhibits/storage/storage_3380.html
4 http://cacm.acm.org/magazines/2009/8/34493-the-pathologies-of-big-data/fulltext#R3
5 http://www.cnet.com/news/facebook-processes-more-than-500-tb-of-data-daily/
6 Intel Peer Research on Big Data Analysis. http://www.intel.com/content/www/us/en/big-data/data-insights-peer-research-report.html
7 https://www.ida.gov.sg/~/media/Files/Infocomm%20Landscape/Technology/TechnologyRoadmap/InternetOfThings.pdf
8 http://mmlab.snu.ac.kr/~mchen/min_paper/Min-0-JNL-4-0-BigData-MONET2013.pdf
9 http://www.techamerica.org/Docs/fileManager.cfm?f=techamerica-bigdatareport-final.pdf
10 http://cacm.acm.org/magazines/2009/8/34493-the-pathologies-of-big-data/fulltext#R3
11 http://bigdatawg.nist.gov/FrontiersInMassiveDataAnalysisPrepub.pdf
12 Couch and Robins, 'Big Data for Defense and Security' *RUSI September 2013*
13 http://blogs.wsj.com/washwire/2013/08/23/nsa-officers-sometimes-spy-on-love-interests/
14 McKinsey Global Institute. May 2011. Big data: The next frontier for innovation, competition, and productivity.
15 http://www.oracle.com/us/products/middleware/data-integration/goldengate11g-realtime-wp-168153.pdf
16 INSA, 'IC ITE: Doing in Common What Is Commonly Done' February 2013
17 INSA, 'IC ITE: Doing in Common What Is Commonly Done' February 2013
18 http://www.fbi.gov/about-us/cjis/n-dex
19 http://ijis.org/docs/NDEX%20Policy%20and%20Operating%20Manual%203.0.pdf
20 http://www.fbi.gov/about-us/cjis/n-dex
21 http://www.dhs.gov/xlibrary/assets/niac/niac-intelligence-information-sharing-final-report-01102012.pdf
22 https://downloads.cloudsecurityalliance.org/initiatives/bdwg/Big_Data_Top_Ten_v1.pdf

23 https://downloads.cloudsecurityalliance.org/initiatives/bdwg/Big_Data_Analytics_for_Security_Intelligence.pdf
24 https://downloads.cloudsecurityalliance.org/initiatives/bdwg/Big_Data_Analytics_for_Security_Intelligence.pdf
25 http://www.mongodb.com/nosql-explained
26 http://www.mongodb.com/nosql-explained
27 http://www.darkreading.com/application-security/database-security/does-nosql-mean-no-security/d/d-id/1136913?
28 Okman et al. 'Security Issues in NoSQL Databases' Trust, Security and Privacy in Computing and Communications, 2011 IEEE 10th International Conference
29 https://www.defcon.org/images/defcon-21/dc-21-presentations/Chow/DEFCON-21-Chow-Abusing-NoSQL-Databases.pdf
30 http://bigdatawg.nist.gov/FrontiersInMassiveDataAnalysisPrepub.pdf
31 http://www.fbi.gov/about-us/cjis/fingerprints_biometrics

13
Strategic Cyber Intelligence: An Examination of Practices across Industry, Government, and Military

Constance Uthoff

Introduction

Over the past decade, despite advances in cyber defense technology, there has been a growth in the sophistication and persistence of cyber attacks and online nefarious activity. Criminals, insiders, and terror groups have had substantial success in their efforts using cyber means. Nations have witnessed the rapid emergence of cyber use as a part of or related to international conflict. The development of advanced, pervasive attacks against the critical infrastructure is an added reminder that cyber attacks hold the potential to destabilize nations, and many states now consider cyberspace a military operational domain. Currently, exhaustive resources are spent on strategies to defend cyberspace, though attacks are often discovered after a network has been compromised.

The escalating success, frequency, and cost of cyber attacks; the length of time adversaries can go undetected in systems; the increased threat of cyber insiders; the mounting cost of cyber defense coupled with the growth in organized cyber crime; as well as an expanding attack surface have all illustrated the need for intelligence-led solutions to address the increasing cyber-security challenges. For decision makers to fully comprehend the threats, to address the rising frequency and cost of attacks and determine how to best allocate resources, a more comprehensive methodology has become necessary. This critical need was reflected in a 2011 report by the Intelligence and National Security Alliance's (INSA) Cyber Council, *Cyber Intelligence: Setting the Landscape for an Emerging Discipline*. Though, currently, there is not a clear assessment of the total impact of cyber attacks, the authors of the report argue that the cost demonstrates the need for a cyber-security process supported by a sophisticated cyber-intelligence methodology.

Similarly, according to Eddie Schwartz, vice president of Global Security Solutions for Verizon Enterprise Solutions, 'Yesterday's approach to fighting cyber attacks in which each enterprise attempts to combat well-funded adversaries alone or build these complex cyber capabilities themselves is just not feasible. The realities of today's threat landscape require enterprises to focus on understanding the business context of an attack, and taking direct remediate action. Reliable and high fidelity cyber intelligence is critical to detecting targeted cyber attacks and to implementing a timely and effective response' (Dark Reading 2014). Many now argue that cyber intelligence should not be compartmentalized from intelligence, but is rather a component of intelligence. In more recent years, a consensus has been established across both public and private sectors that the gathering, processing, and dissemination of cyber intelligence is essential. As a result, it can be helpful to look at the dialogue surrounding the need for cyber intelligence and how it is or may be implemented.

Through a review of current dialogue, reports, and processes related to cyber-intelligence tradecraft, analysis, and application, this chapter will examine definitions and models of cyber intelligence use across industry, the government, and military. Though this chapter only briefly touches on some of these areas, (areas that continue to evolve), it remains that by understanding the application of various cyber intelligence frameworks, decision makers can more fully select strategies that meet their organizational needs and anticipate future threats with greater certainty. These insights can also lead to the development of more effective approaches and the design of supporting technology.

Defining cyber intelligence

Cyber intelligence is not universally defined, though like traditional intelligence, cyber intelligence can mean different things to different consumers. Before examining definitions, it is important to recognize that consumers of cyber intelligence dictate what is of essential value and relevant to their organization or agency. Since each organization is bound by varying mission sets, operational environments, resources, privacy and liability considerations as well as legal constraints, the context of the working environment will be paramount to how that organization or agency will use cyber intelligence and what constraints may exist relative to how (and why) intelligence is gathered and processed.

Understandably, the priorities that are applicable to a corporation are significantly different for military commanders and government leaders. As a result, though many agencies across the private and public space may utilize cyber intelligence for similar reasons – to predict risk, address cyber threats, and provide leaders with actionable intelligence (strategic, operational, and tactical) for decision-making – they each also have unique organizational approaches, objectives, and challenges that can impact and be impacted by the developing tradecraft. The following sections will examine cyber intelligence in the public and private space.

Corporate cyber intelligence

For those seeking to define cyber intelligence in the corporate environment, it can be helpful to first examine why cyber intelligence is important and why there is a growing consensus that companies should adopt a cyber-intelligence tradecraft. Currently, many businesses have a limited ability to detect and respond to sophisticated breaches to their systems, a sobering fact considering that a 2014 Cisco Annual Security Report claims cyber threats and vulnerabilities have reached their highest level since 2000, and the average cost of cyber crime can reach up to $8.6 million annually per company.

The attack against Target cost approximately $148 million to clean up; there was also turnover in senior leadership, and the loss of consumer confidence negatively impacted stock and profits. As if that was not enough for business leaders to consider, additional pressure comes from FTC and SEC claims that companies, board members, and senior leaders can be held responsible for cyber breaches. Commissioner Luis Aguilar of the US Securities and Exchange Commission (SEC) in a 2014 speech stated that boards 'must take seriously their responsibility to ensure that management has implemented effective risk management protocols. There can be little doubt that cyber-risk also must be considered as part of [a] board's overall risk oversight.'

Finally, though detection time is improving, the time an attacker can be in a victim's network is on average six to eight months. (Though there have been breaches that have gone undetected for years.) As a result of these combined threats to their organizations, business leaders are realizing that it is vital to have a more effective and proactive approach to protecting their organizations from cyber intrusions.

Some companies, like IBM, Deloitte, and Lockheed Martin, recognize the real risk to business functions and have already developed cyber-intelligence methodologies to address the expanding challenges. These companies identify correlations among cyber security, data protection, client information, and profit. Targeted cyber attacks can negatively impact operations and client perception, so in order to protect clients, profits, and reputation it is essential for companies to include cyber intelligence as part of their cyber-security apparatus. Cyber intelligence – as a tradecraft – offers a more robust view into the entire operating environment (both internal and external) and examines the cyber threat and risk in the context of what is vitally important from the perspective of the organization. It goes beyond security and allows business leaders to have a better overall concept of costs and risks related to cyber attacks and defense overtime.

Understanding the need for cyber intelligence is an important first step and provides a context for its application, but what is cyber intelligence? A 2013 Intelligence National Security Alliance (INSA) report entitled *Strategic Cyber Intelligence* notes that, though there is not one definition of cyber intelligence, it includes the following:

> the collection and analysis of information that produces timely reporting, with context and relevance to a supported decision maker. The information sources used for cyber intelligence are no more limited than they are for any other field that is observed and analyzed by intelligence professionals.

In relation to this, Dennesen (2014) further reflects that

> it is an analytic discipline relying on information collected from traditional intelligence sources intended to inform decision makers on issues pertaining to operations at all levels of cyber. Relevant data to be analyzed may be about network data, ongoing cyber activity throughout the world or potentially relevant geopolitical events.

Experts from the Cyber Intelligence Research Consortium at the Software Engineering Institute (SEI) at Carnegie Mellon University define cyber intelligence as 'the acquisition and analysis of information to identify, track and predict cyber capabilities, intentions and activities to offer courses of action that enhance decision making' (Townsend 2013). Cyber intelligence, implemented correctly, can enable cyber warning; shape cyber-security efforts; help allocate resources; influence strategy, policy, and doctrine; and inform decision makers.

Cyber intelligence is not universally defined, nor is there a universal approach to the tradecraft. Quite the contrary, there is a wide range of organizational applications of cyber intelligence across the private sector. There are ad-hoc approaches, some lack the depth of a cyber-intelligence tradecraft, some are more technically driven than others, and some look more like cyber-security functions than an intelligence methodology.

For example, Deloitte uses an intelligence-led approach in order to manage cyber risk for their consulting clients. In order to help clients make informed decisions, they assess and provide context for threats specific to an organization. They analyze internal and external sources in order to provide the full breadth of relevant and actionable information. IBM also offers a proactive approach to cyber intelligence, and they examine more than just technical information; they also seek to identify the motivations, capabilities, intentions, objectives, vulnerabilities, and, if possible, identities of threat actors. This extends beyond non-traditional security methods.

Many organizations, like Verizon, CrowdStrike, and Deloitte, have cyber-intelligence centers meant to enable their clients with a deeper comprehension of cyber threats specific to their organization. The new Verizon Cyber Intelligence Center (VCIC) provides clients with cyber-threat intelligence and actionable information. By collecting data from threat management tools and the VERIS cyber incident database, clients can gain insight to threats, ideally before the threats can negatively impact the company.

These examples reflect a growing demand for a more robust and proactive approach to cyber security, but there are other attempts to more fully examine the need, approach, and benefits of cyber intelligence. To this end, the Intelligence and National Security Alliance (INSA) and others have contributed to valuable discourse about cyber intelligence through white papers, analysis, research, blogs, and projects. Starting in 2011, INSA's Cyber Task Force produced a series of reports including *Cyber Intelligence – Setting the Landscape for an Emerging Discipline*. In September 2013, March 2014, and October 2014, they released *The Operational Levels of Cyber Intelligence, Strategic Cyber Intelligence* and *Operational Cyber Intelligence*, respectively. These reports collectively define the need for a cyber-intelligence tradecraft, examine definitions, discuss operational layers of cyber intelligence (strategic, operational, and tactical), and further explore the importance and approach of the strategic and operational layers.

To the INSA authors, cyber intelligence provides actionable information to decision makers on the strategic through tactical levels. *Strategic cyber intelligence* is focused on long-term issues that impact (strategic) decision-making. It is not limited to analysis of technical data but includes a more robust comprehension of threats, threat actors, and the long-term implications to a corporation. Understanding what is imperative to the future success of an organization, legal frameworks, or even geopolitical events can help to enable strategic cyber-intelligence objectives (Dennesen 2014).

According to others, operational cyber intelligence involves comprehending the operational environment, objectives, decisions, trends, approaches, and resources related to the activities of malicious actors. This type of intelligence is often connected to people, how they think and operate. It can also can include other threats like environmental ones, for example. Good operational cyber intelligence can influence cyber-security planning to thwart enemy objectives and to minimize risk.

Great attention has been given to the tactical level of cyber intelligence, often at the expense of strategy. It is important to recognize that tactical intelligence and operations, focused on technical data, are real time and related to attempted or successful cyber attacks.

According to Farnham (2013), the author of *Tools and Standards for Cyber Threat Intelligence Projects,* tactical cyber intelligence involves the tactics, techniques, and procedures (TTPs) of the adversary and also examines indicators of compromise (IOCs). (INSA authors of *Operational Levels of Cyber Intelligence* also include TTPs under Operational Cyber Intelligence.) TTPs can help the practitioner analyze a tactical problem. They can also be used to isolate patterns of behavior, weapons, or approaches utilized by an enemy on a tactical level. Indicators of compromise are behaviors or activities on a system that strongly suggest that a computer intrusion has occurred. According to Farnham, IOCs are one of the most easily actionable types of cyber threat intelligence and are often the focus of standards and tools. After IOCs have been discovered and identified, they can become part of a security apparatus and used for early detection of future attack attempts. 'Some of the most commonly recognized IOCs are IP addresses, domain names, uniform resource locators (URLs) and file hashes' (Farnham 2013). Understanding technical threats is part of protecting systems and can be helpful to decision makers. Tactical intelligence can also help to enable operational and strategic thinking.

Unfortunately, more often than not, companies experience coordination gaps between the strategic and the tactical. Understanding cyber operational planning and design as it relates to corporations could be a valuable next step related to aligning the corporate mission with the strategic, operational, and tactical cyber-intelligence goals. It is also important to recognize that by focusing on only one level of cyber intelligence, the technical/tactical, for example, 'consumers such as C-suite executives, executive managers, and other senior leaders may not be getting the right type of cyber intelligence to efficiently and effectively inform the organization's risk management program' (Dennesen 2014).

So, a few questions arise from that: How can executives ensure that they receive the right level of cyber intelligence? How can the three layers nest effectively so that tactical intelligence efforts clearly align with operational and strategic cyber-intelligence objectives (as well as the corporate mission), while maintaining the freedom of action and resources necessary to properly defend a network? Finally, many propose that risk management is tied into cyber-intelligence efforts. What is the most effective way for cyber intelligence to inform risk management?

To apply cyber intelligence correctly, as mentioned earlier, corporations need an approach that is relevant to their organization and related to what is most important to that company. With that in mind, the authors of the 2014 INSA Strategic Cyber Operations Report recommend that

> each organization would be best served by defining the levels of cyber intelligence for its own unique circumstances, using the following six key criteria: 1. The nature, role, and identity of the consumer; 2. The decisions the consumer will make; 3. The time frame in which the consumer tends to operate; 4. The scope of collection; 5. The characterization of potential adversaries; 6. The level of technical aptitude available for cyber intelligence collection.

Taking the time to examine the six criteria can be a powerful first step for organizations, especially those that have limited experience in this area. Not only will these steps help to determine time frames, resources available, and sophistication of corporate cyber security as well as that of the potential attackers that are unique to an organization, but also it can create an expandable benchmark for current and future planning. Changes are rapid in cyberspace, so planning must be flexible, adaptable, and considered part of an ongoing cyber intelligence process. Having well-organized criteria is a good place to start. These six steps are not exhaustive. Organizations can adapt this initial approach to meet future needs.

Next, according to the white paper, senior leaders should consider intelligence requirements in order to set the framework for strategic planning. This includes (1) Commander's Critical Information Requirements (CCIR): What critical information relative to strategic priorities is needed by senior leadership? (2) Priority Intelligence Requirements (PIR): Linked to essential information about the operating environment or the adversary. (3) Friendly Force Information Requirements (FFIR): This includes the information that leaders need to understand about their own capabilities (Dennesen 2014). These items can help to address uncertainty and can inform decision makers and shape strategy by evaluating what is known and what is unknown/needed. It is important to recognize that the priority of certain information can change over time as the situation/strategy develops. Finally, cyber intelligence and risk management are connected (cyber intelligence feeds risk analysis), and the INSA authors of the Strategic Cyber Intelligence Report recommend that the strategic cyber-intelligence function, within the context of an organization's business model, should support the methods of threat assessment, vulnerability assessment, and impact assessment, as suggested by the National Institute of Standards and Technology (NIST).

These steps will help to shape planning efforts, enable participants to more fully comprehend the operational environment/threat landscape, and shape collection efforts. This will also help executives identify key resources they should protect and develop strategic direction related to scalable, business-driven cyber security. Additionally, these steps will provide participants with a structured approach to identify, prepare for, and respond to breaches across business sectors and offer another way to demonstrate due diligence in the face of potential lawsuits and regulatory response to corporate cyber breaches.

The work of Farnham (2014) outlines the steps to creating a cyber-threat intelligence system that can be used to collect, utilize, and share cyber-threat intelligence. Many would argue that cyber-threat intelligence is not really cyber intelligence because it lacks the methodology that takes data/information and develops it into cyber intelligence. Regardless, the information from cyber-threat intelligence can help to proactively defend against attacks and can be used by decision makers to produce actionable results, so it is included in this discussion. As part of this process, Farnham explores two specific areas important to the cyber-intelligence process: cyber-threat intelligence sources and information-sharing tools. Threat intelligence sources include categories that examine the internal, community, and external threat environment. Internal cyber intelligence is collected from within an organization and can come from firewalls, forensics, intrusion protection systems, and other logs. Community and external collection sources are beyond the network perimeter; they are from outside the company. Community sources such as ISACs are sources with a shared interest, where public and private are not, the latter often pay to join the service (Farnham 2014). Again, cyber-threat intelligence sharing, much like cyber intelligence, will be based on the point of view, resources, and need of the consumer and can provide valued-added information to defend against current and emerging threats. In order to understand the application of cyber intelligence, models like the SEI cyber-intelligence analytical framework can be helpful.

In 2012, the Innovation Center at Carnegie Mellon University's Software Engineering Institute (SEI) was tasked by the Office of the Director of National Intelligence to assess cyber intelligence in the public and private space. The assessment, the Cyber Intelligence Tradecraft Project (CITP), examined public and private uses of cyber intelligence, created a baseline, and evaluated participants against a SEI cyber-intelligence analytical framework. The SEI study discussed distinctions between strategic cyber intelligence and functional intelligence. To these authors, strategic cyber intelligence can help decision makers comprehend the relevance of why they should invest in cyber security and what resources make the most sense. It answers the *who* and *why* related to cyber threats. Functional, or 'technical and tailored,' intelligence answers the *what* and *how* (Townsend 2013). Functional analysis is often done to support cybersecurity efforts; however, the report also indicates that cyber security and cyber intelligence 'feed off each other.'

Cyber-intelligence efforts resonate effectively in relation to cybersecurity practices, but according to their study, companies have not yet experienced consistent success or coordination with these practices on the strategy level. The most successful applications of cyber intelligence are found when organizations maintain situational awareness through ongoing cyber risk assessments, so they can appropriately anticipate, prepare for, and respond to cyber threats and threat actors (Townsend 2013). SEI created an approach that mirrors the intelligence cycle steps: planning, collection, processing, analysis, and dissemination. They refer to their similar five processes as environment, data gathering, functional analysis, strategic analysis, and reporting/feedback. This approach incorporates how technology influences the way analysis is done and uniquely identifies the functions that integrate technology (Townsend 2013).

Specific to cyber-intelligence activities, these efforts create an opportunity to gather and process analysis on both a tactical and strategic level as part of the same initiative, stressing the importance of both in relation to short- and long-term business and cyber-security objectives. Of further value, the report captures some challenges and best practices that may occur in various stages of their cyber-intelligence cycle.

Perhaps to be expected, the SEI analysis discovered that the corporations in their study did 'not adhere to any universal standard for establishing and running a cyber-intelligence program, gathering data, or training analysts to interpret the data and communicate findings and performance measures to leadership. Instead, pockets of excellence exist where organizations excel at cyber intelligence by effectively balancing the need to protect network perimeters with the need to look beyond them for strategic insights' (Townsend 2013).

Other results indicated that there were significant problems with information sharing, supply chain risks, limited resources and/or capability to engage in strategic-level analysis and planning, too much data, data collection without focus, a lack of industry standards to apply to cyber-intelligence training and workforce development, and ineffective communication up the chain of command (Townsend 2013). The study also presented many best practices that can be adopted to enhance capabilities across various sectors. Some companies excelled in creating a cyber-security/cyber-intelligence culture; others managed data collection and analysis effectively. One corporation was able to filter threat data in a manner that helped to feed strategy. Some of the best applications of cyber intelligence came from practitioners who built comprehensive threat profiles that included adversary TTPs and the types of targets they went after. This type of invaluable information illustrates how a cyber-intelligence analytical framework can be utilized to measure the effectiveness of cyber intelligence across industries and how best practices can be used to enhance corporate cyber-intelligence initiatives.

There has been significant attention surrounding cyber-intelligence sharing. Various attempts at legislation, including the Cyber Intelligence Sharing and Protection Act and the Cyber Security Information Sharing Act, illustrate the nationally recognized benefit to appropriate cyber-intelligence sharing. Cyber threat intelligence as an output of cyber intelligence is a fundamental part of a robust cyber-security apparatus, and it can help to inform decision makers about the most recent and serious cyber threats, provide insights into the larger cyber environment, enable more effective defense strategies, and help to allocate resources.

Many critical sectors in the private sphere have information-sharing and analysis centers (ISACs) or cyber-intelligence sharing centers (CISCs) that merge with government and intelligence agencies. Just recently, in light of the attacks against Target and others, the Retail Industry Leaders Association (RILA) launched the Retail Cyber Intelligence Sharing Center (R-CISC). This sharing center allows retailers to share cyber-threat intelligence 'among themselves and, via analysts, with public and private stakeholders, such as the US Department of Homeland Security, US Secret Service and the Federal Bureau of Investigation. The R-CISC will also provide advanced training and education and research resources for retailers' (RILA 2014).

This is just one example of cyber-intelligence sharing. For organizations that want to share cyber threat intelligence or other types of incident data, there are a variety of approaches available. Sharing threat intelligence is both a collection and reporting activity, but this information can be of critical importance to thwarting cyber attacks. In 2012, during high-level DDoS attacks against US financial services, the FS-ISAC enabled effective communication channels to share threat information. Using threat intelligence, they created profiles that they updated throughout the day while attack TTPs changed. They successfully shared this and previous attack information with other financial institutions in order to limit the impact of the DDoS attacks. This example illustrates why cyber threat intelligence and information sharing has become so critical, a key topic in cyber legislation, and considered fundamental to meeting the challenges of future cyber attacks.

In the corporate space, cyber intelligence should support organizational missions, goals, and objectives. In this environment, in order to truly make informed decisions, leaders need to have a deeper comprehension of the cyber threat, the sophistication of the attacker, the methodology, and identification of the source, if possible. It is essential for organizations to comprehend the risks, costs, and the potential counters to a cyber threat focused on their particular industry. It is also key for those decision makers to understand their internal and external cyber environments, especially now that CEO and board members may be held liable for breaches to their systems. Once corporations have a clear indication of where threats and attacks are coming from, that information can be shared to further enable defense strategies. Unfortunately, today, the lack of cyber intelligence standards in the private sector has created gaps in application, limits the ability to measure effectiveness, and can leave corporations wondering if cyber-intelligence practices are the best security investments.

Organizations that appropriately apply cyber-intelligence planning, collection, analysis, and reporting can build a more successful cybersecurity apparatus. To do this, they must have a process to follow. This includes planning and a continual evaluation of their cyber risk in relation to threats and vulnerabilities in both the internal and external environment. Creating a method to gather, analyze, and share cyber intelligence based on what is important to the corporation can help provide a benchmark that can be measured and improved upon. These and other efforts will help corporations anticipate, recognize, and prepare for current and future attacks. Without a more robust strategic approach, US companies will continue to face potential financial and reputational devastation brought on by breaches to their networks.

Cyber intelligence: A look at federal agencies

Unlike the private sector, the US government is poised to cover a breadth of national cyber-security objectives; however, like the private sector, applying a proactive approach to cyber security has taken on increased importance. The 2014 National Intelligence Strategy (NIS) illustrates this shift. In September 2014, the Department of National Intelligence (DNI) released its annual National Intelligence Strategy in order to 'promote integration, mission direction and focus resources across 17 intelligence agencies.' Though it has long been recognized that intelligence in cyberspace can increase the ability of a government to gain insight into foreign cyber capabilities, in 2014, for the first time, the NIS included cyber intelligence as a mission objective to be implemented by the intelligence community.

In the 2009 NIS, the cyber mission was cyber-security-centric, but focused primarily on 'enhancing cyber security both by increasing our ability to detect and attribute adversary cyber activity and by expanding our knowledge of the capabilities, intentions, and cyber vulnerabilities of our adversaries.' Over the ensuing five years, cyber intelligence became a new cyber mission objective, presenting a sophisticated, organized methodology to address the escalation of cyber threats against US interests. According to the 2014 NIS, the new topical mission, cyber intelligence, is defined as the 'collection, processing, analysis, and dissemination of information from all sources of intelligence on foreign actors' cyber programs, intentions, capabilities, research and development, tactics, and operational activities and indicators; their characterization, or insight into the components, structures, use, and vulnerabilities of foreign information systems.'

The intelligence community (IC) has the authorities, the capabilities, resources, and mission set to gather all source cyber intelligence on foreign cyber programs in order to inform national-level leadership and shape policy.

In relation to the private sector, the IC has a different mission set and laws that apply to cyber collection. The IC is focused on national security imperatives and 'advancing national objectives. The private sector is primarily concerned with business strategy; the US government is focused on national strategy. From the 2014 NIS, part of the IC cyber mission recognizes that 'customers increasingly rely upon the IC to provide timely, actionable intelligence and deeper insights into current and potential cyber threats and intentions. The IC also provides needed expertise to defend US Government networks along with other critical communications networks and national infrastructure.'

As part of a national imperative to address national cyber-security issues, and recognizing that national cyber security is a team effort, in March 2013, the US Federal Cyber Security Operations Team agreed to specific national cyber-security roles for the Department of Homeland Security (DHS), the Department of Justice (DoJ)/FBI, and the DoD. The DoJ leads the investigation through prosecution of cyber crimes, and the FBI is responsible for domestic cyber-intelligence activities: the collection, analysis, and dissemination of cyber threat intelligence. DHS is the lead department protecting against domestic cyber-security attacks, and the DoD is responsible for defending the nation from attacks from cyberspace.

In regards to domestic cyber threat intelligence analysis and dissemination and for the purpose of shared situational awareness, it is the responsibility of DHS to communicate information through the National Cyber Security and Communications Integration Center (NCCIC). As part of a proactive effort, the center collaborates with five federal cyber-security centers. These centers are from US CYBERCOM, the Defense Cyber Crime Center, the Intelligence Community – Incidence Response Center, the National Security Agency, and the National Cyber Investigative Joint Task Force. They share and integrate cyber threat intelligence, especially intelligence concerning potential threats of national interest. Cyber threat intelligence that stems from these departments helps to provide foreign cyber threat analysis and facilitate collection and information sharing among other IC elements and agencies. In early 2015, the director of national intelligence called for a new Cyber Intelligence Integration Center in order to more effectively share cyber intelligence.

Another way that DHS shares vital cyber threat information is through Joint Indicator Bulletins. Recently, the DHS and FBI sent out a joint intelligence bulletin related to ISIS using social media to find and target current or former service members. Online social media platforms can help extremists to find out information about their targets, their families, where they live, and their activities. The threat bulletin warned those involved with the military to 'scrub' their profiles.

The FBI also has a significant role related to US cyber-security and intelligence efforts. Part of the DoJ, the FBI is dual hatted, covering both law enforcement and cyber-intelligence functions, which means they cover a gamut of cyber activities from investigation and prosecution to cyber-intelligence collection, processing, and reporting. This can give them initial insights into cyber incidents that could potentially become national security threats. In 2002, to meet the growing challenges related to cyber terrorism and cyber crime, the FBI stood up a Cyber Division. As part of the objectives outlined by the Comprehensive National Cyber Security Initiative (CNCI), the FBI and other partners collaborate in areas related to cyber-intelligence collection, analysis, and information sharing.

The FBI is involved in other areas involving national cyber-intelligence coordination efforts. One example is the National Cyber Investigative Joint Task Force (NCIJTF), which coordinates with 20 other law enforcement and intelligence agencies including NSA, the CIA, and DHA. The Cyber Intelligence Section (CybIS) (of the National Cyber Investigative Joint Task Force) provides analysis and actionable intelligence on areas that overlap in cyberspace, especially those that involve terrorism, crime, and foreign intelligence. The FBI has a variety of ways to collect and disperse cyber threat intelligence in close to real time. Cyber Watch (CyWatch) is a 24-hour command center that receives threat reports and assesses whether the intelligence is actionable and compares it to similar industry incidents. Personnel at Cyber Watch connect with the appropriate intelligence elements or law enforcement agencies to take action if necessary. In order to support federal, state, and local law enforcement and provide them with greater access to information for investigations, in 2012 the FBI announced the development of a new cyber-intelligence and research unit, the Domestic Communications Assistance Center (DCAC). This cyber-intelligence unit will collect information from Internet and wireless communications. Finally, part of an initiative to help private companies, the FBI has created iGuardian. Members can receive support if they are victims of a cyber attack, and they can access cyber intelligence, training, and cyber threat information.

One challenge facing the FBI related to cyber-intelligence gathering and sharing comes when agents discover a breach to a private company before the corporation does. During cyber-intelligence collection activities, agents may travel through the Dark Web, check criminal forums, and look for evidence of criminal activity. There they can find evidence of a wide range of criminal action. In the Deep Web, they may find stolen credit cards, consumer information, or data dumps stolen from companies. Agents are also alerted about crimes through interviews with cyber criminals. Sometimes they can trace these back to the organizations that have been breached. At times, they even have knowledge of who the attackers are and where they came from. When the FBI (or Secret Service) finds evidence of a crime, they will alert the organization that has been breached, but too many companies are finding out after the fact, an indication of the success of today's sophisticated (and malicious) cyber actors. Waiting until after the attack has been discovered is too late. One source estimates that about 40 percent to 50 percent of their customer base have regular conversations with the FBI and other agencies have been warned that they have been breached (Messmer 2014). This again demonstrates why corporations need to have a more proactive approach, one that includes a cyber-intelligence process and mechanism to share timely and relevant cyber threat intelligence.

Unfortunately, even when the FBI discovers a breach, they cannot follow up on all of the cyber intrusions that they find. For example, sometimes attacks come from foreign countries that do not recognize US laws and will not cooperate with US federal authorities. Sometimes it is not in the strategic best interest to confront a cyber attacker because by doing so, US agents may reveal capabilities to the adversary. Agents must also measure whether they want to lose the ability to track a cyber adversary. If an attacker stops one attack or disappears, he or she will most likely show up somewhere else, and it will take additional effort and resources to track them again. Unfortunately, at times cyber criminals will be able to circumnavigate US laws and practices to their benefit; however, the FBI continues to communicate to US businesses about cyber threats and offer advice on protection and mitigation. Following the November 2014 attack against Sony Pictures Entertainment, the FBI sent out a flash bulletin warning corporations about a destructive attack aimed at US businesses. This is the first known destructive cyber attack within the US (Finkle 2014).

Another intelligence element, the CIA, has been involved in sophisticated cyber operations that involve cyber intelligence support and processing. It is rare to get insight into CIA cyber activities, but the following illustration is well known and demonstrates that for nation-states and militaries seeking to have the option to use cyber weapons, cyber intelligence is necessary. The Stuxnet program, known as Operation Olympic Games, was a covert operation with collaboration among the CIA (which normally, under US law, leads covert operations), NSA, and Israel's Unit 8200 aimed at gaining access to the programmable logic controllers at Iran's nuclear enrichment facility in Natanz (Sanger 2012). The facility was air gapped, secure computer networks were physically isolated from the surrounding unsecured networks, so it took a variety of collection methods to gain intelligence about the complex, and it specifically required the development and execution of a computer code that would map and create an electronic blueprint of the facility (Sanger 2012). Reportedly, the National Security Administration (NSA), with Israel's Unit 8200 Intelligence Corps, gathered and analyzed the information about the structure, timing of the spin rate of the centrifuges, and the internal system of the facility at Natanz. The cyber intelligence was used to create a method to infiltrate the facility and disrupt the process leading to uranium enrichment via cyber means. Approximately 1,000 centrifuges were destroyed as a result (Sanger 2012). Clearly this was an operation that included the 'the collection, processing, analysis, and dissemination of information from all sources of intelligence on foreign actors' cyber programs' (Clapper 2014).

Stuxnet introduced the concept of a pre-emptive cyber strike, increased dialogue about the use of cyber weapons, concerns about collateral damage, and the application of the Laws of Armed Conflict in Cyberspace. It also demonstrated US/Israeli cyber capabilities. It was a global demonstration of cyber power, but Stuxnet could also prove to be a catalyst that drives other nations to more fully embrace and develop their cyber weapons capabilities. To that end, on a national and international level, as part of global competition (seen in the arms and space races) cyber intelligence methodologies will become increasingly essential, sophisticated, and integrated more closely with cyber operations.

Though the US government has greater access to sophisticated cyber-intelligence collection, analysis, and dissemination tools and support than the private sector, there are also challenges that can impact national security objectives. The private sector owns a majority of US critical services; however, the US government is charged with the protection of national security interests. Privacy concerns and questions about government reach at times strain relationships between government agencies and the private sector; however, cooperation is imperative. During cyber intelligence collection activities, the FBI has been known to discover a breach before a company knows about it, illustrating the critical need for a comprehensive and proactive approach that flows across the public, private space, one that includes timely, relevant, and accurate cyber threat intelligence sharing.

Military cyber intelligence

Though current dialogue related to cyber intelligence is still emerging, as a concept within the military, cyber intelligence has been around for decades. During that time, military analysts collected foreign news media, chatrooms frequented by threat actors, blogs and video from crisis areas, and commercial imagery to cite just a few applications. It also incorporates the more familiar concept of computer network exploitation (Hurley 2012). Computer network exploitation, however, much like cyber espionage, is not cyber intelligence in and of itself, but is much more closely aligned to cyber collection activities (JP Information Operations, CNE falls under CNO).

In the context of cyber intelligence, it is important to recognize that surveillance and reconnaissance are collection activities that support current and future operations. Reconnaissance missions are launched to gather information about the operating environment or the threat; surveillance relates to the observation of people, locations, or things. They integrate both cyber intelligence and operations and call for collaboration between intelligence and operations (JP-2-0 2012).

A current understanding of how (cyber) intelligence can support cyberspace operations or any military operation can be found in a variety of key publications. The Joint and National Intelligence Support to Military Operations (JP 2-01), Joint Publication 3-13 Information Operations (JP 3-13), and Joint Publication 3-12, Cyberspace Operations (JP 3-12) are good places to start.

It appears that JP 3-12 does not differentiate between cyber intelligence and intelligence, but it does address intelligence support to cyberspace operations. The first reference concerns collection: 'Intelligence collected in cyberspace may come from DoD and/or national-level sources and may serve strategic, operational, or tactical requirements.' The second refers to intelligence operations in and through cyberspace: 'National level intelligence organizations, including major DoD agencies, conduct intelligence activities for national intelligence priorities. This intelligence can support a military commander's planning and preparation.' Today, cyberspace is a military domain and, as a result, there are very significant implications related to cyber-intelligence activities.

The February 2003 National Strategy to Secure Cyberspace clearly states that if the United States is attacked via cyber means, the US government reserves the right to respond 'in an appropriate manner.' The response can involve any element of power, it can be kinetic, or it might involve the use of US cyber weapons designed to attack and disrupt a predetermined enemy target.

For the US military, the 'ability to operate in and from the global commons-space, international waters and airspace, and cyberspace is important to project power anywhere in the world from secure bases of operation' (US National Defense Strategy 2005). To stress the importance of cyberspace and military operations, the US Department of Defense stood up US CYBERCOM in 2010, a sub-unified command of US STRATCOM. The DoD Strategy for Operating in Cyberspace (2011) reinforced that perspective that the DoD must ensure that it has the necessary capabilities to operate effectively in all domains: air, land, maritime, space, and cyberspace. For this to be possible, cyber intelligence may be needed to support both kinetic and/or cyber operations.

Currently, the head of US CYBERCOM also oversees NSA efforts. This relationship illustrates just how closely cyber intelligence and military operations are intertwined. Similarly, out of the 17 intelligence gathering/processing, eight (including NSA) fall under the Department of Defense. There are also several related service components that fall under US CYBERCOM including the US Army Intelligence and Security Command and the 780th Military Intelligence Brigade, for example. Effective intelligence/cyber intelligence is intrinsically important to the success of military operations and national security efforts. With this in mind, this section will briefly look at military examples, but will also discuss NSA and the Snowden disclosures. It will also take a quick look at Presidential Policy Directive 20 and the relationships among offensive cyber operations, the potential preparation of the cyber battle space and cyber intelligence.

Cyber intelligence can help prevent strategic surprise in cyberspace. It can integrate relevant cyber information and threat warning into decision makers' strategic approach and can be used to evaluate adversary actions in cyberspace.

According to the 2012 Navy Cyber Power 2020 Report, 'The Navy effectively evaluates adversary actions in cyberspace through dedicated cyber intelligence collections and analysis and by fully integrating timely and relevant cyber information and threat warnings into the commander's operational picture.' Depending on the mission, cyber information can be collected in and through cyberspace to support operations in a variety of domains (air, sea, land, space, cyberspace) and processed as intelligence. Likewise, all source intelligence collection can be used to support an operation in cyberspace. This would involve multi-INT collections and analyses across the entire intelligence process (cycle). The focus is on the mission, and cyber-intelligence planning and activities would be shaped to support that mission.

Presidential Policy Directive 20 (PPD-20) speaks to some of the intentions of the US government to protect US interests in cyberspace, maintain the ability to collect intelligence, and defend against adversaries that seek to harm US national interests. Originally this was a classified document; however, as a result of the Snowden disclosures, PPD-20 is now available on the Internet. PPD-20 was signed by President Barack Obama in 2012 and discusses cyber operations, which includes cyber collection (part of the cyber intelligence process), defined as follows:

> Operations and related programs or activities conducted by or on behalf of the United States Government, in or through cyberspace, for the primary purpose of collecting intelligence – including information that can be used for future operations from computers, information or communications systems, or networks with the intent to remain undetected. Cyber collection entails accessing a computer, information system, or network without authorization from the owner or operator of that computer, information system, or network or from a party to a communication or by exceeding authorized access. Cyber collection includes those activities essential and inherent to enabling cyber collection such as inhibiting detection or attribution, even if they create cyber effects (PPD-20 2012).

PPD-20 focuses primarily on defensive cyber effects operations (DCEO) and offensive cyber effects operations (OCEO). It recognizes that the US has a robust cyber collection capability; however, it is important to note that many of the items discussed in PPD-20 would not be attainable without significant and ongoing cyber-intelligence planning, collection, processing, and reporting. For example, when considering OCEO, operations intended to produce cyber effects, a cyber-intelligence approach (based on the definition provided in the 2014 National Intelligence Strategy) is necessary.

As mentioned in an earlier section, in order to use OCEO, the US must have significant cyber intelligence about target networks and systems, the potential consequences/collateral damage, the operating systems – anything concerning the related cyber environment including the people, processes, and location. It may also be necessary to have continued and uncontested access to a target considering that OCEO can offer 'unique and unconventional capabilities with little or no warning to the adversary or target' (PPD-20 2012). OCEO capabilities may take time to develop and sustain, which again indicates collaboration with cyber-intelligence functions. Getting access and knowledge about enemy tactical cyber targets and capabilities is a fundamental part of national security efforts and requires continuous cyber intelligence collection, analysis, and reporting. Discovering and maintaining a target takes time, and a target may not always be available.

One responsibility of the US Armed Forces and their lawyers includes a review of weapons and the potential collateral impact related to their use. Cyber weapons have additional challenges related to their use. Cyber-intelligence practitioners will need resources and have specific targets identified and then prepared long before the need to strike. Stuxnet offers a perfect illustration of the importance and sophistication of US cyber intelligence. It took years of planning and cyber-intelligence gathering prior to and during the actual date of launch. Another example was in 2011, with Operation Odyssey Dawn. The United States considered using cyber weapons against Libya but felt there was not enough time. It takes time to perform recon, infiltrate a military target, and strike, especially if it is meant as a surgical strike with minimal collateral damage. Once used, cyber weapons can also be discovered, reverse engineered, and potentially used against US interests, and the decision to use them must consider a myriad of implications.

Additionally, in order to understand the operational environment related to a cyber operation, cyber intelligence is necessary. Cyber intelligence, as well as traditional intelligence, can provide the commander with a variety of assessments and estimates that facilitate understanding the operational environment. In order to provide commanders with a full assessment of the operating environment, one that now includes cyberspace, cyber intelligence can provide decision makers with a greater understanding of the complex layers of cyberspace, which also happen to be mostly privately owned. Defining the operational environment and the impact of the environment and evaluating actors and courses of action are steps that are part of the joint intelligence preparation of the operating environment. The process related to intelligence preparation can easily be applied to cyberspace and could include CCIRs, PIRS, and/or FFIRs. Some areas of interest within the cyber-operating environment could include technology both friendly and adversary, information and communications technology, personal information environment, physical dimension, and the cognitive dimension. This list is, of course, not exhaustive.

Since the head of NSA also leads USCYBERCOM, this section will briefly look at some of NSA's collection programs disclosed by the Snowden documents. The Snowden disclosers were not limited to PPD-20. His release of exorbitant amounts of information concerning US collection activities, though an act of theft and, some argue, treason, provided insight into the breadth and sophistication of US cyber-intelligence collection. Today, the United States has overwhelming cyber collection capabilities, which can greatly enhance its cyber-intelligence functions. Some of the better-documented collection programs came out of NSA and include PRISM, GENIE, and TURBINE (Gellman 2013). According to one of Snowden's classified leaks, PRISM would collect information directly from servers of noted companies like Microsoft, Yahoo, Google, and Facebook. GENIE was an effort that involved hacking into strategically targeted systems in order to control and have continued access to them.

In 2008, more than 21,000 targets had been compromised with projections to infiltrate a total of 85,000 in the following years (Gellman 2013). TURBINE, active since at least July 2010, is an automated system that manages millions of compromised machines for both cyber-intelligence collection and potentially to have access to disrupt or damage if necessary. TURBINE has infected up to 100,000 devices. With TURMOIL, TURBINE taps into computer networks, monitors data traffic, and locates targets through identification of IP addresses, e-mail, or web cookies (Thomson 2014). To access the more difficult targets and computers, NSA has a Tailored Access Operations (TAO) hacking unit. This unit is comprised of specialized and sophisticated hackers who have provided significant and impressive intelligence to the United States.

The United States is not only collecting through its own programs but also receives cyber information from foreign intelligence services. For example, the Snowden leaks indicated that the British Intelligence Government Communications Headquarters (GCHQ) used Tempora, which is known for capturing metadata and intercepting fiber-optic traffic, and shared the findings with NSA. Other countries, like Germany, also sent large amounts of metadata to NSA (MacAskill 2013).

Even though US experts argued that collection was done legally and most, if not all, countries are involved in espionage as part of their own national security efforts, there has been significant backlash from the American public, diplomatic efforts have been strained, and Internet service providers distanced themselves from Washington in an effort to minimize damage to their reputation and reestablish consumer confidence. Trade negotiations between the US and other countries were impacted, and when the US government put pressure on national and foreign press to stop releasing Snowden documents, the credibility of the press was put into question.

Furthermore, there have been significant privacy issue challenges related to US cyber-intelligence collection activities and information sharing with foreign intelligence services. A report by the *New York*

Times in early 2014 illuminates one glaring potential concern for US law firms related to client/attorney privilege. According to the article, the Australian equivalent of the NSA at one time gathered information between an Indonesian firm and the US law firm it was doing business with. The intelligence was then passed on to the NSA. NSA did not collect the information about a US organization, but it did have it in its possession, which had US attorneys concerned about the US government intercepting and reading their privileged communications (Poitras 2014). Even if the collection activity does not come directly from the NSA, the implications have raised questions related to the protection of corporate and legal data from nation-state cyber espionage. At the time, the leader of NSA, General Alexander, responded, 'NSA has afforded and will continue to afford appropriate protection...during its legal foreign intelligence mission in accordance with privacy procedures required by Congress' (Alexander 2014).

Though the US military and intelligence agencies have sophisticated tools and processes to collect, process, analyze, and distribute cyber intelligence, there are still challenges to information collection and sharing that strain public and foreign confidence and, at the same time, as a result of current laws, limit the amount of protection they can offer the US organizations getting hit the hardest by cyber attacks. Snowden disclosures have impacted US legitimacy, diplomatic leverage, and international relations. From another perspective, cyber-intelligence collection capabilities have become so robust on a national level that there is and will continue to be overwhelming amounts of data to sift through. Analyzing, processing, and storing so much data will be a challenge as well. Centers in Utah and Fort Meade will help with cyber-intelligence production, so information can be used most effectively, but even then there will need to be ways to effectively analyze the data (and collaborate with other INTs) so that important information is not lost.

Conclusion

Cyberspace is a domain ripe to support cyber espionage and intelligence activities because it masks the behavior of foreign collectors, makes it possible to collect large amounts of data, and makes attribution very difficult, leaving plenty of room for plausible deniability among nations. Cyber intelligence is still an emerging discipline, but to fill in the gaps, both the US private and public sectors will have to address current challenges including information sharing: getting the right information to the right people at the right time, creating better tools and processes to handle Big Data and verify the reliability of data, creating tools for attribution, addressing supply chain concerns, finding a methodology for decision-making within the limited time that cyberspace offers without relying only on automation, overcoming international collaboration challenges and privacy issues, and developing a more effective tradecraft. Many organizations are off to a good start, but much still needs to be done. To succeed in the cyber domain in 2014 and beyond, cyber intelligence can play a key role in how we prepare for an uncertain future. Technology is changing the way we work, relate to the world, and perceive warfare. It is also expanding a surface to attack corporations and national security interests. Necessary cyber-intelligence methodologies tailored to fit industry, government, and military needs will help to prevent strategic surprise from adversaries that have the potential to cripple the United States and its economy.

References

Ackerman, S. and J. Kaiman (2014). Chinese military officials charged with stealing U.S. data as tensions escalate. The Guardian. New York; Beijing. Retrieved from: http://www.theguardian.com/technology/2014/may/19/us-chinese-military-officials-cyber-espionage.

Acohido, B. (2013). Why the Shamoon virus looms as a destructive threat? Retrieved from USA Today: http://www.usatoday.com/story/cybertruth/2013/05/16/shamoon-cyber-warfare-hackers-anti-american/2166147/.

Adegbite, S., I. Furey, Y. Kadobayashi, R. Martin, D. Rajnovic, G. Reid, T. Rutkowski, G. Schudel and T. Takahashi (2010). Recommendation ITU-T X.1500 [X.cybex], Cyber security information exchange framework. Study Group 17. International Telecommunication Union.

Alberts, D. S. et al. (2001). Understanding Information Age Warfare. Washington, DC: CCRP Publication Series.

Alexander, K. (2007) Warfighting in Cyberspace. Joint Force Quarterly, 3rd Quarter(46), 58–61. Retrieved from: http://oai.dtic.mil/oai/oai?verb=getRecord&metadataPrefix=html&identifier=ADA518148.

Alexander, K. (2014). NSA Response to ABA. Retrieved from American Bar: Http://www.americanbar.org/content/dam/aba/images/abanews/nsa_response_03102014.pdf.

Almeida, A. (2012). Pakistan Cyber Army Hacks into BSNL. Retrieved on 10 November 2012 from: http://tech2.in.com/news/general/pakistan-cyber-army-hacks-into-bsnl/232572.

Alperovitch, D. (2011). Revealed: Operation Shady RAT, McAfee White Paper, Publication No 33000wp_shady-rat_0811. Retrieved from: http://tinyurl.com/3jo2ob2 (www.mcafee.com/us/resources/white-papers/wpoperation-shady-rat.pdf).

American Intelligence (2007). Iran Launches Massive Exercise. Retrieved from Amerian Intelligence: http://americanintelligence.us/iran-launches-massive-exercise/.

Amoroso, E. (2013). Cyber attacks protecting national infrastructure. New York: Elsevier Inc.

Andress, J. and S. Winterfeld (2011). Cyber Warfare: Techniques, Tactics and Tools for Security Practitioners. Waltham, MA: Elsevier Inc.

Ansell, C. and A. Gash (2007). Collaborative Governance in Research and Practice. Journal of Public Administration Research and Theory, 18: 543–571.

Appelbaum, A. (2014). Russia's information warriors are on the march – we must respond. Retrieved on 7 March 2014 from: www.telegraph.co.uk7news/worldnews/europe/russia.

Applegate, S. (2012). The Principle of Maneuver in Cyber Operations. 4th International Conference on Cyber Conflicts (CYCON). Conference proceedings: 1–13.

Applegate, S. D. (2013). The Dawn of Kinetic Cyber. In K. Podins, J. Stinissen, and M. Maybaum (Eds.), Proceedings of the 5th International Conference on Cyber Conflict. Tallinn, Estonia: NATO CCD COE Publications.

Arenas, M., P. Barcelo, L. Libkin, and F. Murlak (2014). Foundations of Data Exchange. Cambridge: Cambridge University Press.

Arguilla, J. (2011). From Blitzkrieg to Bitskrieg: The Military Encounter with Computers. Communications of the ACM, 54(10): 58–65.

Arquilla, J. and D. Ronfeldt (1999). The Advent of Netwar: Analytic Background. Studies in Conflict & Terrorism, 22(3): 193–206.

Ashiyane (2014). Ashiyane. Retrieved from Ashiyane Security Forum: ashiyane.org

Assange, J. (2006). The Curious Origins of Political Hacktivism. Counterpunch. Petrolia, CA. Retrieved from: http://www.counterpunch.org/2006/11/25/the-curious-origins-of-political-hacktivism/.

Awan, A. N. (2007). Radicalization on the Internet? The Virtual Propagation of Jihadist Media and Its Effects. The RUSI Journal, 152(3): 76–81.

Bada, M., S. Creese, M. Goldsmith, C. Mitchell, and E. Phillips (2014). Computer Emergency Response Teams (CERTs). An Overview. University of London: Global Cyber Security Capacity Centre.

Baker, S. (2010). Going Wobbly on Russia's Cybersecurity Disarmament Proposal?, The Volokh Conspiracy. Retrieved from: http://volokh.com/2010/06/06/going-wobbly-on-russias-cybersecurity-disarmament- proposal/?utm_source=feedburner&utm_medium=feed&utm_campaign=Feed:+volokh/mainfeed+(Th e+Volokh+Conspiracy)&utm_content=Google+Reader.

Baker, S. (2010). Skating on Stilts: Why We Aren't Stopping Tomorrow's Terrorism. Stanford University: Hoover Institution Press.

Barkham, J. (2001). Information Warfare and International Law on the Use of Force. New York University International Law and Politics, 34(1): 57–72.

Barzashka, I. (2013). Are cyber-weapons effective?, RUSI Journal, 158(2): 48–56.

Batini, C., C. Cappiello, C. Francalanci, and A. Maurino (2009). Methodologies for data quality assessment and improvement. ACM Computer Survey, 41, 3, Article 16.

Bennet, A. (2014). Verizon Launches New Cyber Intelligence Center. Retrieved from GSN: Government Security News.

Berkeley, III, A., W. Bush, P. Heasley, J. Nicholson, J. Reid, and M. Wallace (2012). Intelligence Information Sharing Final Report and Recommendations. Retrieved on 30 November 2014 from: http://www.dhs.gov/xlibrary/assets/niac/niac-intelligence-information-sharing-final-report-01102012.pdf

Berzins, J. (2014). Russia's New Generation Warfare in Ukraine: Implications for Latvian Defense Policy. Policy paper no 2, 2014. National Defence Academy of Latvia Center for Security and Strategic Research. Retrieved from: www.naa.mil.lv/-/media/NAA/AZPC/Publikacijas/PP%2002-2014.ashx

Bhatotia, P., A. Wieder, R. Rodrigues, U. A. Acar, and R. Pasquini (2011). Incoop: MapReduce for Incremental Computations. Max Planck Institute for Software Systems (MPI-SWS) and Faculdade de Computação - Universidade Federal de Uberlândia (FACOM/UFU).

Birkas, B. and R. Bourgue (2013). EISAS – European Information Sharing and Alerting System Deployment Feasibility Study. European Union Agency for Network and Information Security.

Blair, D. (2009). The National Intelligence Strategy of the United States of America. Director of National Intelligence: Washington, DC.

Bodeau, D. J. and R. Graubart (2011). Cyber Resiliency Engineering Framework. Bedford, MA: Mitre Corporation.

Booton, J., M. Egan, and A. Samson (2012). Bank of America Hit by Cyber Attack. Retrieved from: http://www.foxbusiness.com/industries/2012/09/18/bank-america-website-experiencing-sporadic-outages/#ixzz2ICH7UUJJ

Boutwell, J. and M. T. Klare (2000). A Scourge of Small Arms. Scientific American, 6: 48–53.

Bowman, W. and L. J. Camp (2013). Protecting the Internet from Dictators: Technical and Policy Solutions to Ensure Online Freedoms. The Innovation Journal: The Public Sector Innovation Journal, 18(1): 1–24.

Bright, P. (2011). Independent Iranian Hacker Claims Responsibility for Comodo Hack. Wired News. Retrieved from: http://www.wired.com/2011/03/comodo_hack/

British Broadcasting Corporation News (2011). U.K. cybercrime costs £27bn a year, Government report. Retrieved from: http://www.bbc.co.U.K./news/U.K.-politics-12492309

Bronk, C. (2013). Hack or Attack? Shamoon and the Evolution of Cyber Conflict. Survival, Global Politics and Strategy. The International Institute for Strategic Studies.

Cabinet Office, U.K. Government (2012). 'The U.K. Cyber Security Strategy: Report on progress – Forward Plans'. Retrieved from: https://www.gov.U.K./government/uploads/system/uploads/attachment_data/file/83757/Cyber_Security_Strategy_Forward_Plans_3-Dec-12_1.pdf.

Cabuk, S., C. I. Dalton, K. Eriksson, D. Kuhlmann, H. V. Ramasamy, G. Ramunno, A-R. Sadeghi, M. Schunter, and C. Stüble (2010). Towards automated security policy enforcement in multi-tenant virtual data centers. Journal of Computer Security, 18: 89–121.

Cahill, T., K. Rozinov and C. Mule (2003). Cyber Warfare Peacekeeping, West Point, NY, United States Military Academy, Proceedings of the 2003 IEEE Workshop on Information Assurance (June 2003): 100–106.

Campbell, K., L. Gordon, M. Loeb, and L. Zhou (2003). The economic cost of publicly announced information security breaches: Empirical evidence from the stock market. Journal of Computer Security, 11(3): 431–448.

Canham, B. (2012). Bank Cyber Attacks: Next steps for avoiding website outages and downtime. Retrieved from: http://www.dotcom-monitor.com/blog/index.php/network-services-monitoring/bank-cyber-attacks-2012-ddos-outages/

Carman, D. (2002). Translation and Analysis of the Doctrine of Information Security of the Russian Federation: Mass media and the politics of identity. Pacific Rim Law & Policy Journal Association.

Carr, J. (2012). Inside Cyber Warfare. Mapping the Cyber Underworld, 2nd Edition. Sebastopol, CA: O'Reilly Media Inc.

Carroll, W. (2008). Iranian Cyber Warfare Threat Assessment. Retrieved from Defense Technology: http://defensetech.org/2008/09/23/iranian-cyber-warfare-threat-assessment/

Cascella, R. G. (2008). The Value of Reputation in Peer-to-Peer Networks. Consumer Communications and Networking Conference: 516–520.

Cavusoglu, H., B. Mishra, and S. Raghunathan (2004). The effect of Internet security breach announcements on market value: Capital market reaction for breached firms and Internet security developers. International Journal of Electronic Commerce, 9(1): 69–105.

Cebrowski, A. (1998). Network Centric Warfare: Its Origin and Future. Naval Institute Proceedings: 28–35.

Centre for the Protection of National Infrastructure (2013a). The National Infrastructure. Retrieved from: http://www.cpni.gov.U.K./about/cni/.

Centre for the Protection of National Infrastructure (2013b). Spear Phishing. Retrieved from: http://www.cpni.gov.U.K./documents/publications/2013/2013053-spear-phishing-understanding-the-threat.pdf?epslanguage=en-gb.

Centre for the Protection of National Infrastructure (2013c). Top 20 critical security controls for cyber defence. Retrieved from: http://www.cpni.gov.U.K./advice/cyber/Critical-controls/.

CESG (2013). The National Technical Authority for Information Assurance. Retrieved from: http://www.cesg.gov.U.K./AboutUs/Pages/aboutusindex.aspx.

Chairman of the Joint Chiefs of Staff (2006). The National Military Strategy for Cyberspace Operations. Washington, DC: United States Department of Defense.

Chairman of the Joint Chiefs of Staff (2014). Joint Publication 1-02: Department of Defense Dictionary of Military and Associated Terms. Washington DC: United States Department of Defense.

Chasmar, J. (2014). Syrian Electronic Army hacks Israeli military's account, tweets 'long live Palestine.' Washington Times. Retrieved from: http://www.washingtontimes.com/news/2014/jul/3/syrian-electronic-army-hacks-israeli-militarys-acc/.

Chen, M., S. Mao, and Y. Liu (2014). Big Data: A Survey. Retrieved 30 November 2014 from: http://mmlab.snu.ac.kr/~mchen/min_paper/Min-0-JNL-4-0-BigData-MONET2013.pdf.

Cheshire, C. (2012). Selective Incentives and Generalized Information Exchange. Social Psychology Quarterly, 70(1): 82–100.

China Daily (2014). China Suspends Cyber Working Group Activities with US to Protest Cyber Theft Indictment. Retrieved from: http://usa.chinadaily.com.cn/us/2014-05/20/content_17519286.htm.

Chow, M. (2014). Abusing NoSQL Databases. Retrieved 30 November 2014 from: https://www.defcon.org/images/defcon-21/dc-21-presentations/Chow/DEFCON-21-Chow-Abusing-NoSQL-Databases.pdf.

Cimiano, P., A. Pivk, L. Schmidt-Thieme, and S. Staab (2005). Ontology Learning from Text: Methods, Evaluation and Applications, volume 123, chapter Learning Taxonomic Relations from Heterogeneous Sources of Evidence, IOS Press: 59–73.

Cisco (2014). Cisco Annual Security Report. Cisco.

Clapper, J. R. (2014). The National Intelligence Strategy of the United States of America. Washington, DC: Director of National Intelligence.

Clark, D. (2010). Characterizing cyberspace: past, present and future. Retrieved from: https://projects.csail.mit.edu/ecir/wiki/images/7/77/Clark_Characterizing_cyberspace_1-2r.pdf.

Clarke, R. and R. Knake (2010). CyberWar: The Next Threat to National Security and What to Do About It. New York: HarperCollins.

Clausewitz, C. V. (1976). On War. (M. Howard & P. Paret, Eds.) (Indexed Ed.). Princeton, NJ: Princeton University Press.
Clayton, M. (2013). Chinese cyber attacks hit key U.S. weapons systems. Are they still reliable? Retrieved from: http://www.csmonitor.com/USA/Military/2013/0528/Chinese-cyberattacks-hit-key-U.S.-weapons-systems.-Are-they-still-reliable.
Clemente, D. (2013). The Internet in 2020: Tranquil or Turbulent? Retrieved from CIGI: http://www.cigionline.org/publications/2013/1/internet-2020-tranquil-or-turbulent.
Cloud Security Alliance (2012). Top Ten Big Data Security and Privacy Challenges. Retrieved 30 November 2014 from: https://downloads.cloudsecurityalliance.org/initiatives/bdwg/Big_Data_Top_Ten_v1.pdf.
Cordesman, A. (2007). Iran's Revolutionary Guards, the Al Qudes Force and Other. Retrieved from Center for Strategic and International Studies: http://csis.org/files/media/csis/pubs/070816_cordesman_report.pdf.
Couch, N. and B. Robins (2013). RUSI Report: Big Data for Defense and Security. Retrieved 30 November 2014 from: http://rrnamb.blogspot.com/2013/10/rusi-report-big-data-for-defense-and.html#.VHtwffldV8E.
Council of Europe (2001). Convention on Cybercrime. Budapest: Hungary. Retrieved from: http://conventions.coe.int/Treaty/EN/Treaties/html/185.htm.
Criminal Justice Information Services (2013). Law Enforcement National Data Exchange (N-DEX), Policy and Operating Manual. Retrieved 30 November 2014 from: http://ijis.org/docs/NDEX Policy and Operating Manual 3.0.pdf.
Cyber Intelligence Task Force (2013). Operational Levels of Cyber Intelligence. Arlington, Virginia. Retrieved from: http://csrc.nist.gov/cyberframework/framework_comments/20131213_charles_alsup_insa_part3.pdf.
Dandurand, L. and O. S. Serrano (2013). Towards Improved Cyber Security Information Sharing. Fifth International Conference on Cyber Conflict.
Danyliw, R., J. Meijer, and Y. Demchenko (2007). The Incident Object Description Exchange Format, RFC5070.
Daubert, T. D., A. Roth, T. R. Bertoson, and A. Blair (2013). United States: President Obama's Cyber Security Executive Order to Impact a Wide Range of Business and Industry. Retrieved from: http://www.mondaq.com/unitedstates/x/222598/Data+Protection+Privacy/President+Obamas+Cyber security+Executive+Order+To+Impact+A+Wide+Range+Of+Business+And+Industry.
Defense Technology (2011). Defense Technology. Retrieved from Social Networking Sites - Weapon, Threat and Target: http://defensetech.org/2011/02/28/social-networking-sites-weapon-threat-target/.
Deloitte (2014) Cyber Intelligence Centre. Retrieved from Deloitte: http://www.cyberintelligencecentre.com/.
Dennesen, K. J. (2014). Strategic Cyber Intelligence. Intelligence and National Secuirty Instiitute.
Department of Air Force (2012). Air Force Policy Directive (AFPD) Directive 10-17, Cyberspace Operations.
US Department of Defense (2010). Cyber Command Fact Sheet. Retrieved from: http://tinyurl.com/7exploy (www.defense.gov/home/features/2010/0410_cybersec/docs/CYBERCOM Fact Sheet to replace online version on OCT 13.pdf).
Department of Defense (2012). JP 3-13 Information Operations. Washington, DC.
Department of Defense (2013). JP 2-01, Joint and National Intelligence Support to Military Operations. Washington, DC.

Department of Defense (2013). JP 3-12 Joint Cyberspace Operations. Washington, DC.

Donskov, Y. and O. G. Nikitin (2005). Special Information Operations in Armed Conflicts, Military Thought, 14(3): 2014.

Downes, L. (2012, August). Why Is the UN Trying to Take over the Internet? Forbes. Retrieved from: http://www.forbes.com/sites/larrydownes/2012/08/09/why-the-un-is-trying-to-take-over-the-internet/.

Downey, O., D. Etzioni, S. Soderland, and D. Weld (2004). Learning text patterns for web information extraction and assessment. In Proceedings of the AAAI Workshop on Adaptive Text Extraction and Mining.

Drew, D and D. Snow (1998). Making Strategy: An Introduction to National Security Processes and Problems, Air University Press, Chapter 11: 163–174.

Dreyfus, S. and J. Assange (2012). Underground: Tales of Hacking, Madness and Obsession on the Electronic Frontier (Intl.). United Kingdom: Canongate.

Dunham, K. (2009). Mobile malware attacks and defense. New York: Elsevier Inc.

Dunn, J. (2013). Barclays bank KVM attack plotted by U.K. cybercrime's 'Mr Big', claim police. Retrieved from TechWorld: http://news.techworld.com/security/3470224/barclays-bank-kvm-attack-plotted-by-U.K.-cybercrimes-mr-big-claim-police/.

Dunning, L. A. and R. Kresman (2013) Privacy Preserving Data Sharing with Anonymous ID Assignment. IEEE Transactions on Information Forensics and Security, 8(2): 402–413.

Dylevsky, I. N., S. A. Komov, S. V. Korotkov, S. N. Rodionov, and A. V. Fedorov (2007). Russian Federation Military Policy in the Area of International Information Security: Regional aspect, Moscow Military Thought, 31 March 2007, referred by Carr, J., 27 July 2009.

Egger, S. (1984). A working definition of serial murder and the reduction of linkage blindness. Journal of Police Science and Administration, 12: 348–357.

Ehlin M. and T. Korner (2007). Handbook on Data Quality. Assessment Methods and Tools. EuroStat. European Commission.

Eichensehr, K. (2015). The Cyber-Law of Nation. Georgetown Law Journal, 103: 317–319.

Eijndhoven, D. (2011). Dutch Government to Design Cyber Defense Doctrine. INFOSSEC, February 27.

Eisenhower Memorial Commission (2011). Project Solarium. Retrieved from: http://www.eisenhowermemorial.org/stories/Project-Solarium.htm.

Eisenhower, D. (1953). Minutes of 155th Meeting of NSC, Papers as President, 1953–1961, NSC Series, Box 4.

Eppler, M. J. and J. Mengis (2004). The Concept of Information Overload: A Review of Literature from Organization Science, Accounting, Marketing, MIS, and Related Disciplines. The Information Society, 20(5): 325–344.

Eppler, M. and J. Mengis (2003). A Framework for Information Overload Research in Organizations. Retrieved from Universita Della Svizzera Italia, Lugano Switzerland: www.bul.unisi.ch/cerca/bul/pubblicazioni/com/pdf/wpca0301.pdf

Eriksson, J. and G. Giacomello (2007). Introduction: Closing the gap between international relations theory and studies of digital-age security. In J. Eriksson and G. Giacomello (Eds.), International relations and security in the digital age. New York: Routledge, 1–29.

Etzioni, A. (2011). Cybersecuirty in the Private Sector. Retrieved from Issues in Science and Technology: http://issues.org/28-1/etzioni-2/.

European Parliament (2004). Regulation (EC) No 460/2004: Establishing the European Network and Information Security Agency. Retrieved from: http://tinyurl.com/7qhlyc8 (eur-lex.europa.eu/LexUriServ/LexUriServ.do?uri=CELEX:32004R0460:EN:HTML).

European Union Agency for Network and Information Security (2013). New report on top trends in the first Cyber Threat Landscape by EU's cyber Agency ENISA. Retrieved from: http://www.enisa.europa.eu/media/press-releases/new-report-on-top-trends-in-the-first-cyber-threat-landscape-by-eu2019s-cyber-agency-enisa

Evans, D. (2011). The Internet of Things: How the Next Evolution of the Internet Is Changing Everything. Retrieved from Cisco: http://www.cisco.com/web/about/ac79/docs/innov/IoT_IBSG_0411FINAL.pdf.

Evans, K. and F. Reeder (2010). A Human Capital Crisis in Cyber Security: Technical Proficiency Matters. Washington, DC: Center for Strategic and International Studies.

Evans, P. C. and M. Annunziata (2012). Industrial Internet: Pushing the Boundaries of Minds and Machines. General Electric.

Ezzedeen AL-Qassam Brigades (2013). Cyber attacks cause Israeli companies heavy losses. Retrieved from AL-Qassam: http://www.qassam.ps/news-7167-Cyber_attacks_cause_Israeli_companies_heavy_losses.html.

Farnham, G. (2013). Tools and Standards for Cyber Threat Intelligence Projects. SANS Institute.

Fast, B. M. (2011). Cyber Intelligence: Setting the Landscape for an Emerging Discipline. Intelligence and National Security Alliance.

Federal Bureau of Investigation (2014). Cyber. Retrieved from FBI: http://www.fbi.gov/about-us/ten-years-after-the-fbi-since-9-11/just-the-facts-1/cyber.

Federal Bureau of Investigation (2010). Fingerprints and Other Biometrics. Retrieved 1 December 2014 from: http://www.fbi.gov/about-us/cjis/fingerprints_biometrics.

Federal Bureau of Investigation (2010). National Instant Criminal Background Check Systems. Retrieved 1 December 2014 from: http://www.fbi.gov/about-us/cjis/nics.

Federal Emergency Management Agency (2008). 'Critical Infrastructure and Key Resources Support Annex'. Retrieved from: http://www.fema.gov/pdf/emergency/nrf/nrf-support-cikr.pdf.

Financial Sector-Information Sharing Analysis Center (2012). Operating Rules. Retrieved from: https://www.fsisac.com/sites/default/files/FS-ISAC_Operating Rules_2012.pdf.

Finkle, J. (2014). Russia hacked hundreds of Western, Asian companies: security firm. Reuters. Washington, DC. Retrieved from: http://www.reuters.com/article/2014/01/22/us-russia-cyberespionage-idUSBREA0L07Q20140122.

Finkle, J. (2014). Exclusive: FBI warns of 'destructive' malware in wake of Sony attack. Retrieved from Reuters.com: http://www.reuters.com/article/2014/12/02/us-sony-cyber security-malware-idUSKCN0JF3FE20141202.

Fitzgerald, M. (1996). Russian Views on Information Warfare. December. Washington, DC: Hudson Institute.

Fitzgerald, M. (1994). Russian Views on Electronic Warfare. The growing role of information technology is rapidly lowering the barrier between war and peace. Retrieved from: http://www.nationalstrategies.com.

Floridi, L. and M. Taddeo (2014). The Ethics of Information Warfare. London: Springer.

Fogarty, K. (2011). North Korea steps forward as new cyberwar villain: March DDOS against South Korea may have been dry run for real attack. Retrieved 6 July 2011 from IT World: http://tinyurl.com/6h5a6od.

Folery, P. (2012). Cyber attacks on U.S. banks shows sector's vulnerability, experts say. Retrieved from: http://www.nj.com/business/index.ssf/2012/09/cyber_attacks_on_us_banking_se.html.

Forss, S. (2012). The Russian Operational-Tactical Iskander Missile System. National Defence University. Finland. Working Papers No 42.

France Presse (2012). Oil giant says August cyber attack targeted all Saudi. Retrieved from: http://www.hurriyetdailynews.com/oil-giant-says-august-cyber-attack-targeted-all-saudi.aspx?pageID=238&nid=36446.

Gal-Or, E. and A. Ghose (2005). The Economic Incentives for Sharing Security Information. Information Systems Research, 16(2): 186–208.

Gallagher, S. (2013). Funded Hacktivism or cyber-terrorists, AmEx attackers have big bankroll. Retrieved from: http://arstechnica.com/security/2013/03/funded-hacktivism-or-cyber-terrorists-amex-attackers-have-big-bankroll/.

Johnson, N. B. (2014). Report: Cyber attacks on critical infrastructure jump 383% in 2011. Retrieved from: http://www.federaltimes.com/article/20120703/IT01/307030004/Report-Cyber-attacks-critical-infrastructure-jump-383-2011.

Gaouette, N. (2013). NSA Spying Risks 35 Billion Dollars in U.S. Technology Sales. Retrieved from: http://www.bloomHYPERLINK "http://www.bloomberg.com/news/2013-11-26/nsa-spying-risks-35-billion-in-u-s-technology-sales.html"berg.com/news/2013-11-26/nsa-spying-risks-35-billion-in-u-s-technology-sales.html.

Garamone, J. (2012). Panetta Spells Out DOD Roles in Cyberdefense, American Forces Press Service. Retrieved from: http://www.defense.gov/news/newsarticle.aspx?id=118187.

Gellman, B. and E. Nakashima (2013). U.S. Spy Agencies Mounted 231 Offensive Cyber-Operations in 2011, Washington Post. Retrieved from: http://articles.washingtonpost.com/2013-08-30/world/41620705_1_computer-worm-former-u-s-officials-obama-administration.

Gellman, B. N. (2013). U.S. Spy Agencies Mounted 231 offensive cyber operations in 2011. The Washington Post.

Gerspacher, N. and F. Lemieux (2010). A market-oriented explanation of the expansion of the role of Europol: filling the demand for criminal intelligence through entrepreneurial initiatives. In F. Lemieux (ed.), International Police Cooperation: Emerging Issues, Theory and Practice. Culompton, UK: Willan Publishing, 62–78.

Gertz, B. (2013). U.S., China Talk Cyber Theft at Strategic Dialogue: China Military Urges Controls on Internet. Washington Free Beacon. Retrieved from: http://freebeacon.com/national-security/us-china-talk-cyber-theft-at-strategic-dialogue/.

Gibson, W. (1984). Neuromancer. New York: Ace Books.

Gilbert, H. (2013). Tackling Cyber Security Risk at Financial Institutions. Retrieved from: http://www.securitymanagement.com/news/tackling-cyber security-risk-financial-institutions-0012576.

Gilmour, S. (2014). Policing Crime and Terrorism in Cyberspace: An Overview. The European Review of Organized Crime, 1(1): 143–159.

Glavic, B. (2014). Big Data Provenance: Challenges and Implications for Benchmarking. Lecture Notes in Computer Science, 8163: 72–80.

Global Security (2011). Israel IDF Military. Retrieved from GlobalSecurity: http://www.globalsecurity.org/military/world/israel/general-staff.htm.

Goldman, D. (2012). Massive bank cyber attack planned. Retrieved from: http://money.cnn.com/2012/12/13/technology/security/bank-cyberattack-blitzkrieg/index.html.

Goldsmith, J. (2013). How Cyber Changes the Laws of War. European Journal of International Law, 24(1): 129–138.

Goldsmith, J. (2010). Cybersecurity Treaties: A Skeptical View. In P. Berkowitz (ed.), Future Challenges in National Security Law. Stanford University: Hoover Institution Press.

Goldsmith, J. (2010). The New Vulnerability. The New Republic. Retrieved from: www.tnr.com/article/books-and-arts/75262/the-new-vulnerability.

Gorman, S. (2013) NSA Officers Spy on Love Interests. Retrieved 30 November 2014 from: http://blogs.wsj.com/washwire/2013/08/23/nsa-officers-sometimes-spy-on-love-interests/.

Government Accountability Office (2013). National Strategy, Roles, and Responsibilities Need to Be Better Defined and More Effectively Implemented. Washington, DC.

Government of UK (2013a). Inside Government. The national security strategy – a strong Britain in an age of uncertainty. Retrieved from: https://www.gov.U.K./government/publications/the-national-security-strategy-a-strong-britain-in-an-age-of-uncertainty.

Government of UK (2013b). Inside Government. Government launches information sharing partnership on cyber security. Retrieved 25 July 2013 from: https://www.gov.U.K./government/news/government-launches-information-sharing-partnership-on-cyber-security.

Government of UK (2013c). Inside Government. Defence Partnership tackles cyber security risks. Retrieved from: https://www.gov.U.K./government/news/defence-partnership-tackles-cyber-security-risks.

Grau, L-W. and Thomas, T. (1996). A Russian View of Future War: Theory and direction, Journal of Slavic Military Studies, 9.3 (September): 501–518.

Grossman, A. and D. Yadron (2014, July 11). U.S. Accuses Chinese Executive of Hacking to Mine Military Data. The Wall Street Journal. Washington, DC. Retrieved from: http://online.wsj.com/articles/u-s-accuses-chinese-executive-of-hacking-to-find-military-data-1405105264.

Gyllenhaal, L. and J. von Braun (2013). Ryska elitförband. Spetsnaz, Osnaz, VDV och andra elitstyrkor [Russian elite units]. Fischer & Co.

Hale, G. (2013). Key SCADA Security Questions for CEOs. Retrieved from: http://www.datacenterjournal.com/it/modernizing-physical-security-and-incorporating-best-practices-into-new-assets/.

Halliday, J. (2011). Anonymous claims to have Stuxnet access. Retrieved 11 February 2011 from guardian.co.UK.: http://www.guardian.co.U.K./technology/2011/feb/14/anonymous-stuxnet-nuclear-iran.

Harris, P. (2013). Chinese army hackers are the tip of the cyber warfare iceberg. Retrieved from: http://www.guardian.co.U.K./technology/2013/feb/23/mandiant-unit-61398-china-hacking.

Healey, J. (ed.). (2013). A Fierce Domain: Conflict in Cyberspace, 1986–2012. Washington, DC: Cyber Conflict Studies Association.

Heickerö, R. (2013). The Dark Sides of the Internet. On Information Warfare and Cyber Threats. Frankfurt am Main: Peter Lang Verlag.

Heickerö, R. (2010). Emerging Cyber Threats and Russian Views on Information Warfare and Information Operations. User Report. Swedish Defence Research Agency. FOI–R–2970–SE.

Herzog, S. (2011). Revisiting the Estonian Cyber Attacks: Digital Threats and Multinational Responses. Journal of Strategic Study, 4(2): 49–60.

Hienz, J. (2011). Water Utilities' SCADA Systems Proven Vulnerable to Cyber Attack. Retrieved from: http://www.defensemedianetwork.com/stories/water-utilities-scada-systems-proven-vulnerable-to-cyber-attack/.

Higginbotham, S. (2013). CES 2013: Connected devices and the Internet of Things. Bloomberg Businessweek Technology.

Hildreth, S. A. (2001). Cyber warfare. Congressional Research Services. The Library of Congress.

Himma, K. E. (2008). Ethical Issues Involving Computer Security: Hacking, Hacktivism, and Counter Hacking. In K. E. Himma and H. T. Tavani (eds.), The Handbook of Information and Computer Ethics. Hoboken, NJ: Wiley & Sons.

Hoffman, D. (2008). KGB Comes in from the Cold, Washington Post, 8 December 2008. Quoted in Carman, D. (2002). Translation and Analysis.

Hofmann, F. G. (2007). Conflict in the 21st Century: The Rise of Hybrid Wars. Arlington, VA: Potomac Institute.

Hopkins, N. (2011). New Stuxnet worm targets companies in Europe. Retrieved from: http://www.theguardian.com/technology/2011/oct/19/stuxnet-worm-europe-duqu?INTCMP=SRCH.

Hunker, J. (2010a). Cyber Power and Cyber War. Issues for NATO Doctrine. NATO Defense College, No. 62: 1–12.

Hunker, J. (2010b). U.S. International Policy for Cyber Security: Five Issues That Won't Go Away. Journal of National Security Law and Policy, 197: 197–215.

Hurley, C. M. (2012). For and from Cyberspace: Conceptualizing Cyber Intelligence, Surveillance, and Reconnaissance. Air Space and Power, 26(6): 12–33.

IBM (2014). Advanced Cyber Threat Intelligence. Retrieved from: www.ibm.com.

IBM (2014). IBM 3380 direct access storage device. Retrieved 30 November 2014 from: http://www--03.ibm.com/ibm/history/exhibits/storage/storage_3380.html.

Inan, A., M. Kantarcioglu, G. Ghinita, and E. Bertino (2012). A hybrid approach to record matching. IEEE Trans. Dependable Sec. Comp., 9(5): 684–698.

Information Warfare Monitor (2009). Tracking Ghostnet: Investigating a Cyber-espionage Network. San Francisco, CA: The SecDev Group.

InfoSecurity (2009). 'Grey Goose 2 Ties Kremlin More Closely to Georgia Cyber-attacks'. Retrieved 20 March 2009 from: http://www.infosecurity-magazine.com/view/762/grey-gosse-2-ties-kremlin-more-closely-to-georgia-cyber.htm.

INSA (2013). IC ITE – Doing in Common What Is Commonly Done. Retrieved 30 November 2014 from: http://www.insaonline.org/i/d/a/Resources/ICITE_Doing.aspx.

Institut Francais d'Analyse Strategique (2012). Structe of Iran's Cyber Warfare. Retrieved from Institut Francais d'Analyse Strategique: http://www.strato analyse.org/fr/spip.php?article223.

Intel IT Center (2012). Big Data Research: IT Manager Survey on Big Data. Retrieved 30 November 2014 from: http://www.intel.com/content/www/us/en/big-data/data-insights-peer-research-report.html.

International Telecommunications Union of the United Nations (2012). FAQs on FLAME. Retrieved from: http://www.itu.int/cyber security/Articles/FAQs_on_FLAME.pdf.

Iran Media Program (2013). The Supreme Council of Cyberspace: Centralizing Internet Governance in Iran. Retrieved from: http://iranmediaresearch.org/en/blog/227/13/04/08/1323#sthash.edPlBJYg.dpuf.

Jacobs, A. (2009). Communications of the ACM. Retrieved 30 November 2014 from: http://cacm.acm.org/magazines/2009/8/34493-the-pathologies-of-big-data/fulltext#R3.

Jaycox, M. M. (2012). The Cyber Security Act was a surveillance bill in disguise. Retrieved from: http://www.guardian.co.U.K./commentisfree/2012/aug/02/cyber security-act-surveillance-bill-disguise.

Jensen, E. T. (2012). Cyber Attacks: Proportionality and Precautions in Attack. Journal of International Law Studies, 89: 198–219.

Jones, B. (2014). Russia to open electronic warfare base on Finland's border. IHS Jane's 360. Retrieved from: http://www.HYPERLINK "http://www.janes.com/article/35635/russia-to-open-electronic-warfare-base-on-finland-s-border"janes.com/article/35635/russia-to-open-electronic-warfare-base-on-finland-s-border.

Kahn, H. (2007). On Thermonuclear War. Piscataway, NJ: Transactions Publishers.

Kaiman, J. (2014). China reacts furiously to U.S. cyber-espionage charges. The Guardian. Beijing. Retrieved from: http://www.theguardian.com/world/2014/may/20/china-reacts-furiously-us-cyber-espionage-charges.

Karatzogianni, A. (2013). A Cyberconflict Analysis of the 2011 Arab Spring Uprisings. In G. Youngs (ed.), Digital World: Connectivity, Creativity and Rights. Abingdon, Oxon: Routledge, 0–21.

Kaspersky Labs (2014). IT Security Risks Survey 2014: A Business Approach to Managing Data Security Threats. Retrieved from: http://media.kaspersky.com/en/IT_Security_Risks_Survey_2014_Global_report.pdf.

Kaspersky Labs' Global Research & Analysis Team. (2014). The Epic Turla Operation. Retrieved from: https://securelist.com/analysis/publications/65545/the-epic-turla-operation/.

Kavhaz Center (2013). Russia launches massive cyber terrorist attack on Lithuania. Retrieved from: http://www.kavkazcenter.com/eng/content/2013/05/28/17832.shtml.

Keil, T. (2014). Cyber Security and Critical Infrastructure Protection (CIP). Retrieved from: http://www.npstc.org/cyberCIP.jsp.

Kelley, O. (2008). Cyberspace Domain: A Warfighting Substantiated Operational Environment Imperative. United States Army War College. Retrieved from: http://oai.dtic.mil/oai/oai?verb=getRecord&metadataPrefix=html&identifier=ADA479775.

Khalamayzer (2012). Protecting the Grid: Risk Management in Electric Utilities. Retrieved from: http://www.propertycasualty360.com/2012/09/17/protecting-the-grid-risk-management-in-electric-ut.

Kharrazi, M. (2012). CE 817 Advanced Network Security 817-902 Lecture. Tehran, Iran.

Kharrazi, M. (2014). Advanced Network Security CE 40-817. Tehran, Iran.

Kharrazi, M. (2014). CE 817: Advanced Network Security Homework 2. Tehran, Iran.
Kijewski, P. and P. Pawlinski (2013). Proactive Detection and Automated Exchange of Network Security Incidents. NATO. Retrieved from: http://www.cert.pl/PDF/MP-IST-111-18.pdf.
Kozel, R. (2010). Stare Decisis as Judicial Doctrine. Washington and Lee Law Review.
Krapp, P. (2005). Terror and Play, or What Was Hacktivism? Grey Room, 21: 70–93.
Kukashkin, A. N. and A. I. Yefimov (1995). The Security of the Infosphere of Strategic Defense System. Military Thoughts, 5.
Kumar, M. (2012). Cyber Attacks on Six Major American Banks. Retrieved from: http://thehackernews.com/2012/10/cyber-attacks-on-six-major-american.html.
Kuperman, G. J. (2011). Health-information exchange: why are we doing it, and what are we doing? Journal of American Medicine Informatics Association, 18: 678–682.
Kurian, J. (2012). Indian Cyber Army hacks Pak websites. Deccan Chronical. Retrieved from: www.deccanchronicle.com/channels/nation/south/?indian-cyber-army?-hacks-pak-websites-203
Kushner, D. (2013). The Real Story of Stuxnet. Retrieved from: http://spectrum.ieee.org/telecom/security/the-real-story-of-stuxnet.
Landau, S. (2014). Under the Radar: NSA's Efforts to Secure Private-Sector Telecommunications Infrastructure. Journal of National Security Law & Policy, 4(1): 411–429.
Langner, R. (2011). Stuxnet: Dissecting a Cyberwarfare Weapon. IEEE Security & Privacy Magazine, 9(3): 49–51.
Laurén, A-L. (2014). Putin's man warns of Swedish 'Russophobia'. Retrieved from: www.svd.se/nyheter/utrikes/putins-man-varnar-for-svenskt rysshat_3640114.svd?sidan=1.
Lebowitz, M. (2011). China military admits cyber warfare unit exists. Retrieved from Security News: www.msnbc.msn.com/id/43189050/ns/technology_and_science-security/t/ china-military-admits-cyberwarfare-unit-exists.
Leijonhielm, J., J. Hedenskog, J. Knoph, I. Oldberg, W. Unge, and C. Vendil (2000). 'Rysk militär förmåga i ett tioårsperpektiv. En förnyad bedömning 2000'. Användarrapport FOA-R-01758-17--SE (Stockholm, FOA). [Russian military capability in a ten-year perspective]
Lemieux, F. (2010). International Police Cooperation: Emerging Issues, Theory and Practice. Cullompton, UK: Willan Publishing,
Lemos, R. (2014, September 4). In case of cyber attack: NATO members ready to pledge mutual defense. Ars Technica. Newport. Retrieved from: http://arstechnica.com/security/2014/09/in-case-of-cyberattack-nato-members-ready-to-pledge-mutual-defense/.
Lewis, J. (2008). Securing Cyberspace for the 44th Presidency: A Report of the CSIS Commission on Cyber Security for the 44th Presidency. Retrieved from CSIS: http://csis.org/files/media/csis/pubs/081208_securingcyberspace_44.pdf.
Lewis, J. (2014). To Regulate or Not to Regulate Cyber Security: That Is the Question. RSA Conference, 26 February 2014, San Francisco, California, United States.

Lewis, J. A. (2011). Cyber Security and Cyberwarfare. Washington, DC: Center for Strategic and International Studies.

Li, X. (2011). Factors influencing the willingness to contribute information to online communities. New Media & Society, 13(2): 279–296.

Liang, Q. and W. Xiangsui (1999). Unrestricted Warfare. Beijing: PLA Literature and Arts Publishing House.

Libicki, M., D. Senty, and J. Pollack (2014). Hackers Wanted: An Examination of the Cyber Security Labor Market. Washington, DC: Rand Corporation.

Liddell-Hart, B. H. (1967). Strategy: The Indirect Approach, 2nd revised ed. London: Faber and Faber Limited, 430.

Limno, A. N. and M. F. Krysanov (2003). Information Warfare and Camouflage, Concealment and Deception, Military Thought, 12(2): 213–223.

Lin, H. S. (2010). Offensive Cyber Operation and the Use of Force. National Security Law and Policy, 63: 63–80.

Lin, P., F. Allhof, and K. Abney (2014). Is Warfare the Right Frame for the Cyber Debate? In L. Floridi and M. Taddeo (eds.), The Ethics of Information Warfare. London: Springer, 39–60.

Liu, C. Z., H. Zafar, and Y. A. Au (2014). Rethinking FS-ISAC: An IT Security Information Sharing Network Model for the Financial Services Sector, Communications of the Association for Information Systems: Vol. 34, Article 2.

Lotan, G., E. Graeff, M. Ananny, D. Gaffney, I. Pearce, and D. Boyd (2011). The Arab Spring: The Revolutions Were Tweeted: Information Flows during the 2011 Tunisian and Egyptian Revolutions, International Journal of Communication, 5: 1375–1405.

Lotrionte, C. (2015). Countering State-Sponsored Cyber Economic Espionage Under International Law, 40 North Carolina Journal of International Law and Commercial Regulation, 40: 443–463.

Lucas, G. R. (2014). NSA Management Directive #424: Secrecy and Privacy in the Aftermath of Edward Snowden. Ethics and International Affairs, 28(1): 29–38.

Luiijf, E., K. Besseling, and P. de Graaf (2013). Nineteen National Cyber Security Strategy. International Journal of Critical Infrastructure, 9(1): 3–31.

Lynn, W. (2011). Remarks on the Department of Defense Cyber Strategy. As delivered by Deputy Secretary of Defense William J. Lynn, III, National Defense University, Washington, DC, 14 July 2011. Retrieved from: http://www.defense.gov/speeches/speech.aspx?speechid=1593

MacAskill, E. (2013). GCHQ Taps Fiber Optic Cable for Secret Access to World Communications. Retrieved from The Guardian: http://www.theguardian.com/U.K./2013/jun/21/gchq-cables-secret-world-communications-nsa.

Magnuson, S. (2013). Growing Black Market for Cyber-Attack Tools Scares Senior DoD Official. Retrieved from: http://www.nationaldefensemagazine.org/blog/Lists/Posts/Post.aspx?ID=1064.

Malkin, T. (2013, March). Secure Computation for Big Data - Springer. Retrieved 30 November 2014 from: http://link.springer.com/chapter/10.1007/978-3-642-36594-2_20.

Malkin, T. (2013). Secure Computation for Big Data. Theory of Cryptography. Berlin: Springer.

Mandiant (2013). APT1: Exposing One of China's Cyber Espionage Units. Alexandria, VA. Retrieved from: http://intelreport.mandiant.com/.

Mandiant (2014). 2014 Threat Report. Mandiant: A FireEye Company.

Mansfield-Devine, S. (2011). Anonymous: serious threat or mere annoyance? Network Security, 11(1): 4–10.
Manyaka, J., M. Chui, B. Brown, J. Bughin, R. Dobbs, C. Roxburgh, and A. Byers (2011). Big data: The next frontier for innovation, competition, and production. Retrieved 30 November 2014 from N-DEX: http://www.fbi.gov/about-us/cjis/n-dex.
Marchibroda J. M. (2007). Health information exchange policy and evaluation. J Biomed Inform, 40(6): S11–S6.
Markoff, J. and T. Shanker (2009). Halted '03 Iraq Plan Illustrates U.S. Fear of Cyberwar Risk. New York Times. Retrieved from: http://www.nytimes.com/2009/08/02/us/politics/02cyber.html?_r=0.
Maslen, G. (2013). Open Doors - Foreign Students Flock to America. Retrieved from University World News: http://www.universityworldnews.com/article.php?story=20131114055159659.
Matrosov, A. et al. (2011). Stuxnet Under the Microscope. ESET Software, Revision 1.31, January 2011. Retrieved from: http://www.eset.com/us/documentation/white-papers.
McAfee (2014) McAfee Labs 2014 Threats Predictions. Retrieved from: http://www.mcafee.com/us/resources/reports/rp-threats-predictions-2014.pdf.
McAfee Corporation and Center for Stategic & International Studies (2013). The Economic Impact of Cybercrime and Cyber Espionage. Santa Clara, CA. Retrieved from: http://www.mcafee.com/us/resources/reports/rp-economic-impact-cybercrime.pdf.
McKenzie, K. (2001). The Rise of Asymmetric Threats: Priorities for Defense Planning, Quadrennial Defense Review, Chapter 3: 75–105.
McTaggart, C. (2011). Hadoop/MapReduce. Retrieved 1 December 2014 from: http://www.cs.colorado.edu/~kena/classes/5448/s11/presentations/hadoop.pdf.
Menn, J. (2014). Exclusive: NSA infiltrated RSA security more deeply than thought – study. Retrieved from Reuters: http://www.reuters.com/article/2014/03/31/us-usa-security-nsa-rsa-idUSBREA2U0TY20140331.
Messmer, E. (2014). How do the FBI and Secret Service know your network has been breached before you do. Retrieved from Network World: http://www.networkworld.com/article/2175582/security/how-do-the-fbi-and-secret-service-know-your-network-has-been-breached-before-you-do-.html.
Microsoft (2014). Cyberspace 2025. Retrieved from Navigating the Future of Cyber Security Policy: http://www.microsoft.com/security/cyber security/cyberspace2025/#chapter-1.
Mite, V. (2007, May 30). Estonia: Attacks Seen as First Case of Cyberwar. Radio Free Europe/Radio Liberty. Retrieved from: http://www.rferl.org/content/article/1076805.html.
Mitre (2013a). Taxii: An overview. Retrieved from: http://taxii.mitre.org/about/documents/TAXII_Overview_briefing_July_2013.pdf.
Mitre (2013b). Use cases (STIX). Retrieved from: http://stix.mitre.org/language/usecases.html.
Mitre (2013c). Cyber observable expression. Retrieved from: http://cybox.mitre.org/.
Molander, R., A. Riddle, and P. Wilson (1996). Strategic Information Warfare, A New Face of War, National Defense Research Institute, RAND.

MongoDB (2014). NoSQL Databases Explained. Retrieved 30 November 2014 from: http://www.mongodb.com/nosql-explained.

Moriarty, K. (2010). Real-time Inter-network Defense (RID), RFC 6045.

Moriarty, K. and B. Trammell (2010). Transport of Real-Time Inter-Network Defense (RID) Messages, RFC 6046.

Muncaster, P. (2014). Second Pro-Government Hacking Group 'Syrian Malware Team' Uncovered. Retrieved from InfoSecurity Magazine: http://www.infosecurity-magazine.com/news/government-hacking-syrian-malware/.

Murry, T. (2012). Banks can't prevent cyber attacks like those hitting PNC, Key, U.S. Bank this week. Retrieved from: http://www.cleveland.com/business/index.ssf/2012/09/banks_cant_prevent_cyber_attac.html.

Nakashima, E. (2011). List of cyber-weapons developed by Pentagon to streamline computer warfare. Retrieved from Washington Post: http://www.washingtonpost.com/national/list-of-cyber-weapons-developed-by-pentagon-to-streamline-computer-warfare/2011/05/31/AGSublFH_story.html.

Napolitano, J. (2012). Testimony of DHS Secretary Janet Napolitano before the Senate Committee on Homeland Security and Governmental Affairs for a hearing entitled Securing America's Future: The Cyber Security Act of 2012. Retrieved from the Department of Homeland Security: http://www.dhs.gov/news/2012/02/16/testimony-dhs-secretary-janet-napolitano-senate-committee-homeland-security-and.

National Cyber Bureau. (2011). Advancing National Cyberspace Capabilities. Retrieved from National Cyber Bureau: http://www.pmo.gov.il/English/PrimeMinistersOffice/DivisionsAndAuthorities/cyber/Documents/Advancing%20National%20Cyberspace%20Capabilities.pdf.

National Cyber Bureau. (2011). Mission of the Bureau. Retrieved from Mission of the Bureau: http://www.pmo.gov.il/English/PrimeMinistersOffice/DivisionsAndAuthorities/cyber/Pages/default.aspx.

National Institute Standards and Technology (2013). Frontiers in Massive Data Analysis. Retrieved 30 November 2014 from: http://bigdatawg.nist.gov/FrontiersInMassiveDataAnalysisPrepub.pdf.

National Intelligence Program (2012). FY 2013 Congressional Budget Justification Volume I National Intelligence Program Summary. Washington, DC: National Intelligence.

National Research Council (2013). Professionalizing the Nation's Cyber Security Workforce? Criteria for Decision-Making. Washington, DC: National Academy of Sciences.

National Security Council (2009). The Comprehensive National Cyber Security Initiative, US Government. Retrieved from: http://www.whitehouse.gov/cyber security/comprehensive-national-cyber security-initiative.

NATO (2014). Protecting critical infrastructure. Retrieved from: http://www.nato.int/cps/en/SID-3D171718-0F4FED96/natolive/news_92793.htm?selectedLocale=en.

NATO (2014). NATO Wales Summit Guide. Retrieved from: http://www.nato.int/cps/en/natohq/news_112107.htm?selectedLocale=en.

NATO Cooperative Cyber Defence Centre of Excellence (2013). Tallinn Manual on the International Law Applicable to Cyber Warfare. (M. N. Schmitt, ed.). Tallinn, Estonia: Cambridge University Press, 1–215.

Neale, M. (2000). No Maps for These Territories. United States: Docurama.

Neer, T. and M. E. O'Toole (2014). The Violence of the Islamic State of Syria (ISIS): A Behavioral Perspective. Violence and Gender, 1(4): 145–156.

New Oxford American Dictionary (2005). Second Edition, Oxford University Press.

NRC (2009) Technology, Policy, Law, and Ethics Regarding U.S. Acquisition and Use of Cyber Attack Capabilities. Washington, DC: National Academy Press.

Okman, L., N. Gal-Oz, Y. Gonan, E. Gudes, and J. Abramov (2011). 2011 IEEE 10th International Conference on Trust, Security and Privacy in Computing and Communications. Piscataway: IEEE, 541–547.

Oracle GoldenGate 11g (2012). Real-Time Access to Real-Time Information. Retrieved 1 December 2014 from: http://www.oracle.com/us/products/middleware/data-integration/goldengate11g-realtime-wp-168153.pdf.

Orman, H. (2003). The Morris Worm: A Fifteen-Year Perspective. IEEE Security and Privacy: 35–43.

O'Toole, T. (2014). FTC's Unfairness Authority Upheld In Wyndham Data Security Litigation. Privacy and Security Report.

Paganini, P. (2014). SNAKE - The last cyber espionage campaign on a large scale. Retrieved from: http://securityaffairs.co/wordpress/22875/intelligence/snake-cyber-espionage-campaign.html.

Parajon, C. (2014). An International Law Response to Economic Cyber Espionage. Connecticut Law Review, 46: 1165–1169.

Parks, R. and D. Duggan (2001). Principles of Cyber-warfare. United States Military Academy, Proceedings of the 2001 IEEE Workshop on Information Assurance and Security, West Point, NY, June 5–6, 2001: 122–125.

Parrish, K. (2011). Cyber Threat Grows More Destructive, American Forces Press Service, July 15.

Perez, A. (2012). U.S. Banks Hit by Cyber-Attacks. Retrieved from: http://www.theepochtimes.com/n2/business/us-banks-hit-by-cyber-attacks-297587.html.

Perlroth, N. (2012). In Cyberattack on Saudi Firm, U.S. Sees Iran Firing Back. Retrieved from: http://www.nytimes.com/2012/10/24/business/global/cyberattack-on-saudi-oil-firm-disquiets-us.html?pagewanted=all&_r=0.

Peterson, A. (2013). How Iranian Hackers Used the Cloud to Attack Major Banks and Why It Matters. Retrieved from ThinkProgress: http://thinkprogress.org/security/2013/01/09/1424171/bank-hackings-iran-botnets-cloud/.

Phythian, M. (2013). Understanding the Intelligence Cycle. New York: Routledge.

Pino, R., A. Kott, and M. Shevenell (2014). Cyber Security Systems for Human Cognition Augmentation. New York: Springer.

Pirumov, V. (1996). Nekotorye aspekty informatsionnoi voiny. Conference speech in Moscow, May 1996.

Poitras, L. (2014). Spying by N.S.A. Ally Entangled U.S. Law Firm. Retrieved from New York Times: http://www.nytimes.com/2014/02/16/us/eavesdropping-ensnared-american-law-firm.html?_r=0.

Ponemon Institute (2014). 2014 Cost of Data Breach Study: United States. Retrieved from: http://www.accudatasystems.com/assets/2014-cost-of-a-data-breach-study.pdf.

Ponemon Institute (2012). A Study of Retail Banks & DDoS Attacks Report. Retrieved from: http://www.corero.com/resources/files/analyst reports/CNS_Report_Ponemon_Jan13.pdf.

Ramakrsihnan, R. and J. Gehrke (2002). Database Management Systems. New York: McGraw-Hill.

Rass, S. and D. Slamanig (2013). Cryptography for security and privacy in cloud computing. Boston: Artech House.

Rastorguyev, S. G. (1998). Informatsionnoi Voiny (Information warfare). Radio i Svjaz, referred in Thomas, T. (2004) Russian and Chinese Information Warfare: Theory and Practise. Foreign Military Studies Office, Fort Leavenworth. PowerPoint, June 1998.

Rayman, N. (2014). The World's Top 5 Cyber Crime Hotspots. Retrieved from Time Magazine: http://time.com/3087768/the-worlds-5-cybercrime-hotspots/.

Reading, D. (2014). New Verizon Cyber Intelligence Center Helps Speed Detection, Mitigation of Cyberthreats for Enterprises, Government Agencies.

Reeve, A. (2013). Big Data and NoSQL: The Problem with Relational Databases - InFocus. Retrieved 1 December 2014 from: https://infocus.emc.com/april_reeve/big-data-and-nosql-the-problem-with-relational-databases/.

Reuters (2014). New NSA Chief Vows More Transparency Embattled Agency. Retrieved from International Business Times: http://www.ibtimes.com/new-nsa-chief-vows-more-transparency-embattled-agency-1583401.

Rid, T. (2012). Cyber War Will Not Take Place. Journal of Strategic Studies, 35(1): 5–32.

RILA (2014). Retailers Launch Comprehensive Cyber Intelligence Sharing Center. Retrieved from: http://www.rila.org/news/topnews/Pages/RetailersLaunchComprehensiveCyberIntelligenceSharingCenter.aspx.

Rothman, P. (2012). Cyber terror rages in the banking sector. Retrieved from: http://www.securityinfowatch.com/blog/10796084/cyber-terror-rages-in-the-banking-sector.

Ruiz-Casado, M., E. Alfonseca, and P. Castells (2007). Automatising the learning of lexical patterns: An application to the enrichment of Wordnet by extracting semantic relationships from Wikipedia. Data and Knowledge Engineering, 61: 484–499.

Rutkowski, A., Y. Kadobayashi, I. Furey, D. Rajnovic, R. Martin, and T. Takahashi (2010). Cybex – The Cyber Security Information Exchange Network. ACM Computer Communication Review, 40(5): 59–64.

Ruus, K. (2008). Cyber War I: Estonia Attacked from Russia. European Affairs, 9: 1.

Sagar, P. R. (2010). Pakistani cyber army hack CBI's website. Retrieved from: http://www.dnaindia.com/india/report_pakistani-cyber-army-hack-cbi-s-website_1476703.

Sanger, D. (2012). Confront and Conceal. New York: Random House.

Sanger, D. (2012). Obama Order Sped Up Wave of Cyberattacks Against Iran. Retrieved from The New York Times: http://www.espm.br/rjclipping/2012/junho/47425.pdf.

Sanger, D. E. and J. Markoff (2010). After Google's Stand on China, U.S. Treads Lightly. Retrieved from The New York Times: http://www.nytimes.com/2010/01/15/world/asia/15diplo.html?ref=technology.

Scheoff, M. (1998). Cybercrime, Cyberterrorism, Cyberwarfare, Averting an Electronic Waterloo. Washington, DC: Center for Strategic and International Studies.

Scheptycki, J. (2004). Organizational Pathologies in Police Intelligence Systems: Some Contributions to the Lexicon of Intelligence-led Policing. European Journal of Criminology, 1(3): 307–332.

Schifrin, N. (2009). Mumbai Terror Attacks: 7 Pakistanis Charged. Retrieved from: http://abcnews.go.com/International/mumbai-terror-attacks-pakistanis-charged/story?id=9176592#.UKvnqYfon4T.

Schmidt, N. (2014). Critical Comments on Current Research Agenda in Cyber Security. Defence and Strategy, 1: 29–38.

Schmitt, M. (2013). Tallinn Manual on the International Law Applicable to Cyber Warfare. Cambridge: Cambridge University Press.

Schmitt, M. (2010). Operations in International Law: The Use of Force, Collective Security, Self-Defense, and Armed Conflicts. In Cyber Computer Science and Telecommunications Board (ed.), Deterring Cyberattacks: Informing Strategies and Developing Options for U.S. Policy. Washington, DC: National Academy Press, 151–178.

Schmitt, M. (1999). Computer Network Attack and Use of Force in International Law: Thoughts on a Normative Framework. Columbia Journal of Transnational Law, 37(4): 890–934.

Schmitt, M. and L. Vihul (2014). Proxy Wars in Cyber Space: The Evolving International Law of Attribution. Fletcher Security Review, 1(2): 55–73.

Seabrook, J. (2013). Network Insecurity: Are We Losing the Battle Against Cybercrime. Retrieved from The New Yorker: http://www.newyorker.com/magazine/2013/05/20/network-insecurity.

Seacord, R. C., D. Plakosh, and G. A. Lewis (2003). Modernizing Legacy Systems: Software Technologies, Engineering Processes, and Business Practices. Boston: Addison-Wesley Professional.

Segal, A. (2012). Understanding China's Cyber Policy. Retrieved from The Diplomat: http://the-diplomat.com/china-power/2012/04/07/understanding-chinas-cyber-policy/.

Seredynski, M., P. Bouvry, and M. A. Klopotek (2007). Modeling the Evolution of Cooperative Behavior in Ad Hoc Networks Using a Game Based Model. Computation Intelligence and Games: 96–103.

Serookiy, Y. (2004). Psychological-Information Warfare: Lessons of Afghanistan, Military Thought, 13(1).

Shakarian, P., J. Shakarian, and A. Ruef (2013). Introduction to Cyber-Warfare: A Multidisciplinary Approach. Waltham, MA: Elsevier Inc.

Sharif University (2014). Fifth Annual Intrusion and Defend Competition in Cyberspace. Retrieved from CERT Sharif University: https://cert.sharif.edu/?a=contentNews.id&id=128 and https://cert.sharif.edu/ctf.

Singh, S. and J. Krupakar (2014). Indo–US Cooperation in Countering Cyber Terrorism: Challenges and Limitations. Strategic Analysis, 38(5): 703–716.

SKYNews (2013). Cyber Threat: Spies and Big Firms Join Forces. Retrieved from: http://news.sky.com/story/1070111/cyber-threat-spies-and-big-firms-join-forces.

Small Arms Survey (2003). Graduate Institute of International and Development Studies, Geneva, Switzerland.

Smith, J. P. (2006). Developing a Reliable Methodology for Computer Network Operations Threat of Iran. Retrieved from Federation of American Scientists: http://fas.org/irp/eprint/cno-iran.pdf.

Snyder, T. (2014) The Haze of Propaganda. Retrieved 1 March 2014 from: www.nybooks.com/blogs/nyrblog/2014/mar/01.

Sokov, N. (2007). The Origins and Prospects for Russian Nuclear Doctrine. The Non Proliferation Review, 14(2): 207–226.

Stenersen, A. (2008). The Internet: A virtual training camp? Terrorism and Political Violence, 20: 215–233.

Stephenson, P. and K. Gilbert (2013). Investigating Computer-Related Crime. New York: CRC Press.

Steward, P. (2014). U.S., Australia to add cyber realm to defense treaty. Reuters. San Francisco, California. Retrieved from: http://www.reuters.com/article/2011/09/15/us-usa-cyber-australia-idUSTRE78E05I20110915.

Stoll, C. (1989). The Cuckoo's Egg: Tracking a Spy Through the Maze of Computer Espionage. New York: Pocket Books.

Stoll, C. (2005). The Cuckoo's Egg: Tracking a Spy Through the Maze of Computer Espionage. New York: Doubleday.

Stratecast (2014). The Hidden Truth Behind Shadow IT. Six Trends Impacting Your Security Posture. Frost & Sullivan. Retrieved from: http://www.mcafee.com/us/resources/reports/rp-six-trends-security.pdf.

Sullivan, G. and J. Dubik (1994). War in the Information Age. Carlisle Barracks, PA: U.S. Army War College.

Sultana, S., M. Shehab, E. Bertino (2013). Secure provenance transmission for streaming data. IEEE Transaction Knowledge Data Engineering, 25(8): 1890–1903.

Sutton, M. (2013). ME govts lag on cyber protection for utilities. Retrieved from: http://www.itp.net/593666-me-govts-lag-on-cyber-protection-for-utilities#.UcpCd5DD_Dc.

Symantec Corporation (2014). Symantec's Internet Security Threat Assessment Report. Volume 19. Retrieved from: http://www.symantec.com/content/en/us/enterprise/other_resources/b-istr_main_report_v19_21291018.en-us.pdf.

Symantec Corporation (2014) Stuxnet 0.5: How It Evolved. Retrieved from: http://www.symantec.com/connect/blogs/stuxnet-05-how-it-evolved.

Symantec Corporation (2011). Norton Study Calculates Cost of Global Cybercrime: $114 Billion Annually, Press Release. Retrieved from: http://www.symantec.com/about/news/release/article.jsp?prid=20110907_02.

Syrian Electronin Army (2014). The programming unit in the Syrian Electronic Army. Retrieved from Syrian Electronic Army official website: http://sea.sy/index/en.

Sztipanovits, J. and S. Ying (2013). Foundations for Innovation: Strategic R & D Opportunities for 21st Century Cyber-Physical Systems: Connecting computer and information systems in the physical world. National Institute of Standards and Technology, Washington, DC.

Tam, D. (2012). Facebook processes more than 500 TB of data daily. CNET. Retrieved 30 November 2014 from: http://www.cnet.com/news/facebook-processes-more-than-500-tb-of-data-daily/.

Tavani, H. T. and F. S. Grodzinsky (2014). Trust, betrayal, and whistle blowing: reflections on the Edward Snowden case. ACM SIGCAS Computers and Society, 43(3): 8–13.

The Economist (2010). The threat from the Internet, cyberwar, it is time for countries to start talking about arms control on the Internet. Retrieved from: http://www.economist.com.

The Express Tribune (2010). 36 government sites hacked by Indian Cyber Army. Retrieved 20 November 2012 from: http://tribune.com.pk/story/83967/36-government-websites-hacked-by-indian-cyber-army/.

The Federal Bureau of Investigation (2014). Five Chinese Military Hackers Charged with Cyber Espionage Against U.S. Retrieved 14 September 2014 from: http://www.fbi.gov/news/news_blog/five-chinese-military-hackers-charged-with-cyber-espionage-against-u.s.

The Gawker Media Group (2014). Cyber-terrorist attack on U.S. electrical grid could be gravest short term threat to national security. Retrieved from: http://io9.com/5807192/cyber+terrorist-attack-on-us-electrical-grid-could-be-gravest-short-term-threat-to-national-security.

The Honorable James M. Simon (2012). Toward a General Understanding of Cyber Safety. In M. S. Timothy Sample, #CyberDoc No Borders - No Boundaries National Doctrine for the Cyber Era. Arlington, VA: Potomac Institute Press, 46.

The Russian Government (2010). The Military Doctrine of the Russian Federation. Approved by Russian Federation presidential edict on 5 February 2010. Retrieved from: http://carnegieendowment.org/files/2010russia_military_doctrine.pdf.

The White House (2015). Securing Cyberspace - President Obama Announces New Cyber Security Legislative Proposal and Other Cyber Security Efforts. Washington, DC: Office of the Press Secretary.

The White House (2013). Remarks by the President in the State of the Union Address. Retrieved from Whitehouse.gov: http://www.whitehouse.gov/the-press-office/2013/02/12/remarks-president-state-union-address.

The White House. (2013). Presidential Policy Directive 20. Washington, DC: U.S. Government.

The White House (2012). Fact Sheet: Big Data Across the Federal Government. Retrieved 13 November 2014 from: http://www.whitehouse.gov/sites/default/files/microsites/ostp/big_data_fact_sheet_final.pdf.

The White House (2011). International Strategy for Cyberspace, Washington, DC. Retrieved from: http://tinyurl.com/3aovtx5 (www.whitehouse.gov/sites/default/files/rss_viewer/international_strategy_for_cyberspace.pdf).

The White House (2008). National Security Presidential Directive 54/Homeland Security Presidential Directive 23 (NSPD-54/HSPD-23). Washington, DC, United States.

The White House (2003). The National Strategy to Secure Cyberspace. Washington, DC, U.S. Retrieved from: http://www.dhs.gov/xlibrary/assets/National_Cyberspace_Strategy.pdf.

Thomas, T. (1998). Dialectical versus Empirical Thinking: Ten key elements of Russian understanding of information operations. FMSO Special Study Center For Army Lesson Learned. Fort Leavenworth, KS 66027–1327.

Thomas, T. (2003). Manipulating the Mass Consciousness: Russian & Chechen information war. Tactics in the second Chechen–Russian conflict, 14 April 2003.

Thomas, T. (2004). Russian and Chinese Information Warfare: Theory and Practice. Foreign Military Studies Offices. Fort Leavenworth. PowerPoint, June 2004.

Thomson, I. (2014). NSA's TURBINE robot can pump 'malware into MILLIONS of PCs'. Retrieved from The Register: http://www.theregister.co.U.K./2014/03/12/snowden_docs_show_nsas_malware_turbine_can_pump_out_millions_of_malware_attacks/.

Thornton, R. (2007). Asymmetric warfare: Threat and response in the twenty-first century. Cambridge, UK: Polity Press.

Tikk, E. (2010). Global Cyber Security–Thinking About the Niche for NATO. Review of International Affairs, 30(2): 105–119.

Tikk, E., K. Kaska, and L. Vihul (2008). Cyber Attacks Against Georgia: Legal Lessons Identified. Cooperative Cyber Defence Centre of Excellence. Retrieved from: https://ccdcoe.org/publications/books/legalconsiderations.pdf.

Tiller, E. and F. Cross (2006). What is legal doctrine? Northwestern University Law Review, 100:1.

Tirenin, W. and D. Faatz (1999). A Concept for Strategic Cyber Defense, IEEE Military Communications Conference, Atlantic City, NJ.

TNN (2012). India's transmission, distribution sectors most vulnerable: Blackout report. Retrieved from: http://articles.economictimes.indiatimes.com/2012-08-30/news/33499615_1_grid-collapse-cyber-attacks-power-systems.

Townsend, T. M. (2013). SEI Emerging Technology Center: Cyber Intelligence Tradecraft Project Summary of Key Findings. Software Engineer Institute: Carnegie Melon.

Treadstone 7 (2014). Syrian Electronic Army Exposure Post 2 - Op Israel. Retrieved from The Cyber Shafarat: http://cybershafarat.com/2014/09/02/syrian-electronic-army-exposure-post-2-op-israel-june-1-2013-uncovered-turkish-alignment/.

Tsymbal, V. I. (1995). Kontseptsiya Informatsionnoi Voiny. Lecture at the Russian–U.S. conference on Evolving Post Cold War National Security Issues, Moscow, 12–14 September 1995.

US Department of Defense (2010). Cyber Command Fact Sheet. Washington, DC.

US Joint Chief of Staff (2014). Information Operations. Joint Publication 3-13. Retrieved from: http://www.fas.org/irp/doddir/dod/jp3_13.pdf.

US STRATCOM (2009). The Cyber Warfare Lexicon. Washington, DC: Department of Defense.

U.S.-CERT (2013). United States Computer Emergency Readiness Team. Retrieved from: http://www.us-cert.gov/.

United States Army Training and Doctrine Command (2010). TRADOC Pamphlet 525-7-8 The United States Army's Cyberspace Operations Concept Capability Plan 2016–2028. Washington, DC: Department of the Army.

United States of America National Senate (2012). Myth vs. Fact: The Cyber Security Act of 2012, Democratic Policy and Communications Centre. Retrieved from: http://www.dpc.senate.gov/docs/fs-112-2-179.pdf.

Valeriano, B. and R. Maness (2014). The dynamics of cyber conflict between rival antagonists, 2001–11. Journal of Peace Research, 51(3): 347–360.

Vendil Pallin, C. and F. Westerlund (2010). Russia's Military Doctrine – Expected News. RUFS Briefing no. 3, February 2010. FOI.

Vermesan, O., P. Friess, P. Guillemin, S. Gusmeroli, H. Sundmaeker, and A. Bassi (2009). Internet of things strategic research roadmap, Internet of Things. Global Technological and Societal Trends.

Vest, J. R. and L. D. Gamm (2010). Health information exchange: persistent challenges and new strategies. Journal of American Medicine Informatics Association, 17: 288–294.

Warden, J. (1995). The Enemy as a System, Airpower Journal (Spring 1995).

Watt, N. (2013). David Cameron Challenges China to Be More Open About Cyber-Security, Guardian. Retrieved from: www.theguardian.com/politics/2013/dec/04/david-cameron-challenges-china-cyber-security.

Watts, S. (2011). Proposal for cyber war rules of engagement. Retrieved 5 March 2013 from: http://news.bbc.co.U.K./1/hi/programmes/newsnight/9386445.stm.

Waxman, M. (2011). Cyber-Attacks and the Use of Force: Back to the Future of Article 2(4). Yale Journal of International Law, 36(2): 421–458.

Webster's New World Dictionary and Thesaurus (2002). Second Edition. New York: Wiley and Sons.

Wigmore, I. and M. Rouse (2014). Internet of Things (IoT). Retrieved 14 September 2014 from: http://whatis.techtarget.com/definition/Internet-of-Things.

Williams, M. G. (2013). Armed Forces Communication and Electronics Association (AFCEA). Conference Proceedings from the 4th Annual Cyber Security Symposium, Cyber Security Symposium 2013, Washington, DC.

World Economic Forum (WEF) (2014). Insight Report: Global Risks 2014. Retrieved from World Economic Forum: http://www3.weforum.org/docs/WEF_GlobalRisks_Report_2014.pdf.

Yost, D. S. (2010). NATO's evolving purposes and the next Strategic Concept. International Affairs, 86(2): 489–522.

Index

9781137455543